D0385550

£1·41

539.7
N2134 A

Committee on Atomic and Molecular Physics
NATIONAL RESEARCH COUNCIL

NATIONAL ACADEMY OF SCIENCES
WASHINGTON, D.C.
1971

NOTICE: The study reported herein was undertaken under the aegis of the National Research Council, with the express approval of the Governing Board of the NRC. Such approval indicated that the Board considered that the problem is of national significance; that elucidation of the problem required scientific or technical competence; and that the resources of NRC were particularly suitable to the conduct of the project. The institutional responsibilities of the NRC were then discharged in the following manner:

The members of the study committee were selected for their individual scholarly competence and judgment with due consideration for the balance and breadth of disciplines. Responsibility for all aspects of this report rests with the study committee, to whom sincere appreciation is expressed.

Although the reports of study committees are not submitted for approval to the Academy membership or to the Council, each report is reviewed by a second group of scientists according to procedures established and monitored by the Academy's Report Review Committee. Such reviews are intended to determine, *inter alia*, whether the major questions and relevant points of view have been addressed and whether the reported findings, conclusions, and recommendations arose from the available data and information. Distribution of the report is permitted only after satisfactory completion of this review process.

Library
I.U.P.
Indiana, Pa.
539.7 N2134a
C.1

Library of Congress Catalog Card Number 75-177419

ISBN 0-309-01931-1

Available from

Printing and Publishing Office
National Academy of Sciences
2101 Constitution Avenue
Washington, D.C. 20418

Printed in the United States of America

Preface

The purpose of this report is to describe activities and current problems in the field of atomic and molecular physics, and it is hoped that such description, including some outline of trends and needs, will be helpful to administrators of science. Another purpose of the study is to acquaint scientists in other fields, students, and perhaps nonscientists too, with the content of atomic and molecular physics and its contributions to knowledge and application.

As indicated in the introductory chapter, the study was initiated several years ago, and since then some fifty scientists have assisted in its conduct. The report was prepared under the guidance of the Committee on Atomic and Molecular Physics of the National Research Council. The material was organized and prepared in its final form by B. Bederson, L. T. Crane, and S. J. Smith; major contributions to the report also were made by S. Bashkin, A. Dalgarno, F. M. Pipkin, and F. T. Smith; the survey described in this report was conducted by R. Geballe and the results analyzed by L. T. Crane. Others who contributed much to the study included F. Allen, A. P. Altshuller, J. C. Baird, C. F. Barnett, E. Bauer, L. M. Branscomb, S. Brodsky, J. Callaway, J. Cooper, S. Datz, E. Davidson, S. Drell, B. Donnaly, M. A. Fineman, W. L. Fite, R. Geballe, A. E. S. Greene, E. F. Greene, D.

Halford, J. H. Hall, K. Kessler, L. J. Kieffer, J. G. King, H. Mark, E. A. Mason, E. W. McDaniel, M. McDermott, J. W. McGowan, A. V. Phelps, P. H. Rose, A. L. Schawlow, G. Schultz, J. A. Simpson, R. Sinclair, S. J. Smith, and H. H. Stroke.

The Committee acknowledges appreciation to the National Science Foundation for its support of the publication of the study. The Committee is also indebted to the NRC Division of Physical Sciences for its staff support, including editorial assistance provided by Miss Bertita E. Compton and Mrs. Jacqueline Boraks.

COMMITTEE ON ATOMIC
AND MOLECULAR PHYSICS

Benjamin Bederson, *Chairman*
Stanley Bashkin
Nicolaas Bloembergen
Lewis M. Branscomb
Langdon T. Crane
Alec Dalgarno
Ronald Geballe
A. Javan
Felix T. Smith
Stephen J. Smith
G. King Walters
Richard Zare

Contents

I ATOMIC AND MOLECULAR PHYSICS RESEARCH AND RECOMMENDED POLICIES FOR ITS SUPPORT

1. Summary 3
2. Introduction 7

A. Background of the Report, 7
B. Atomic and Molecular Physics: Its Nature and Significance, 9

3. Statistical Survey 11

A. Scope, 11
B. Definition and Procedure, 12
C. Results and Interpretation, 14

4. Conclusions 20

A. Background, 20
B. Recommendations, 25

APPENDIX I.A, 37

II TECHNICAL REVIEW OF ATOMIC AND MOLECULAR PHYSICS

5. Introduction 43

A. The Physics of the Gaseous State, 43
B. A Frontier for Relevant Research, 46

6. Outline History of Atomic and Molecular Physics 50
7. Atomic and Molecular Physics as a Basic Science 60

A. Fundamental Atomic and Molecular Physics, 60
B. Simple Atomic Systems, 64
C. Complex Atoms and Molecules, 65
D. Calculation of Atomic and Molecular Properties, 68
E. Atomic Collisions, 70

8. Relation of Atomic and Molecular Physics to Other Branches of Physics 73

A. Nuclear Physics, 73
B. Solid-State Physics, 77
C. Plasma Physics, 82
D. Plasma Diagnostics–Spectroscopy, 85
E. Astrophysics, 85

9. Relation of Atomic and Molecular Physics to Other Sciences and Education 90

A. Chemistry, 91
B. Aeronomy, 98
C. Biology, 102
D. Education, 104

10. Lasers and Quantum Electronics 108

A. Atomic and Molecular Physics in Laser Development, 108
B. Quantum Electronics, 111
C. The Use of Laser Methods to Study Atomic Physics Properties and Systems, 112
D. Saturated Absorption and Ultrahigh-Resolution Spectroscopy, 116

11. Atomic and Molecular Physics in Technology 118

A. Fusion Research, 118
B. Atomic and Molecular Physics in Space and Defense Technology, 122
C. Missile and Space Vehicle Re-entry, 124
D. Combustion, 124
E. Nuclear Fireball, 125
F. High-Temperature Technology, 125
G. Air Pollution Research, 125
H. Electric Discharges, 127
I. Surface and Emission Properties, 127
J. High-Vacuum Technology, 129

12. Technological Workhorses: Metrology's Atomic Standards 130

A. Broadcasting, 133
B. Communications, 133
C. Power Distribution, 133
D. Lasers, 134
E. Navigation, 134
F. Space, 135
G. Astronomy, 135

I ATOMIC AND MOLECULAR PHYSICS RESEARCH AND RECOMMENDED POLICIES FOR ITS SUPPORT

1 Summary

Atomic and molecular physics research has consistently presented exciting new problems to physicists. Fifty years ago the difficulties of explaining atomic spectra resulted in the formulation of the quantum theory. Today atomic and molecular scientists are making unique contributions to such problems as pollution, energetics of the atmosphere, and the life sciences. In the physical sciences, atomic and molecular physics clearly is among the leaders in its broad applicability to problems of recognized national importance.

In recent years, new techniques for examining some of the very-low-energy atomic and molecular processes that underlie atmospheric and life processes have been developed. With these techniques, scientists are now examining the dynamics of highly excited and highly ionized atoms and molecules, such as those that occur in space and stellar matter. As a result of these advances, atomic and molecular scientists are undertaking research programs that are far beyond the scope of classical physics. Formerly, an atomic and molecular scientist who became involved in atmospheric- or life-science problems soon found that he was isolated from his professional colleagues, primarily because his scientific field was not yet prepared to contribute meaningfully to such complex problems. Fortunately, that era is ending. Atomic and molecular research now provides a strong connecting link between the physical sciences and other fields and subfields of science; the greatest opportunities for young atomic and molecular scientists lie in multidisciplinary research areas.

The major technical advances that underlie this broadening of scope and capability did not come about because of a conscious effort of the federal government to promote atomic and molecular physics research. In sharp contrast to the careful and articulate attention given to improving fundamental knowledge in most of the other subfields of physics, the potential of atomic and molecular physics was not recognized, and it therefore was not fostered in government agency programs in fundamental science. However, atomic and molecular physics did receive support in connection with a great many developmental programs in defense and space research. From these joint efforts in basic and applied science came many of the research tools that are now of such great use in multidisciplinary efforts.

For nearly two decades, atomic and molecular physicists were in the frustrating position of having many potentially productive problems to attack and inadequate means with which to attack them. Older experimental methods were not applicable to the much more sophisticated investigations that were required. New and more powerful techniques began to emerge in the late 1950's and early 1960's. The development of these techniques coincided with growing needs in a wide variety of federal agencies for greater knowledge about the atmospheres, about new methods of power production, and about space sciences in general. Although support from the National Science Foundation (NSF) remained approximately level, the Atomic Energy Commission (AEC), the U.S. Air Force, and the National Aeronautics and Space Administration (NASA) increased their support to keep pace with the growth in experimental competence of physicists working in the general area of atomic and molecular physics and the related fields of plasma physics and space science.

The picture changed radically after 1966. Unofficial NSF statistics show that federal support for the combined fields of atomic and molecular physics and plasma physics* decreased by approximately 28 percent between 1966 and 1970. The survey conducted in connec-

*Support programs within the federal government generally have responsibility for both atomic and molecular physics and plasma physics. The close interrelationship between these two subfields often necessitates their being considered together.

tion with this report revealed a drop of 20 percent in federal support for atomic and molecular physics between 1964 and 1968. This support curtailment was precipitous and uncoordinated, leaving some of the most productive scientists in atomic and molecular physics either without support or with only a small fraction of their former support. Never having established a serious support effort in atomic and molecular physics, the NSF was unable to meet more than a small fraction of the demand for support produced by the funding cuts of other agencies and the continuing growth of activity and interest in this field.

Because of the particular importance of atomic and molecular physics research to a multitude of immediate problems of great national concern, the Committee on Atomic and Molecular Physics developed the following recommendations based on the findings of their survey and their assessment of the history and contributions of atomic and molecular physics. We believe that the policies outlined in these six recommendations are necessary if this field is to continue to produce at a level comparable with that of other basic science disciplines. (The thinking that underlies these recommendations and a brief discussion of each are presented in Chapter 4.)

RECOMMENDATION 1 Federal support of basic atomic and molecular research in universities should be increased by at least $8.2 million in the next three to five years to ensure that current levels of activity will not be reduced.*

RECOMMENDATION 2 Atomic and molecular science has received significant stimulation from its close association with and support through applied research programs in the federal government. Current support reductions in many of these programs have jeopardized their effectiveness and that of atomic and molecular physics research. Therefore, the federal government should make a concerted attempt to revitalize the support of basic research through applied programs in the National Aeronautics and Space Administration, the Atomic

*As later sections of Part I indicate, the survey conducted by the Committee dealt with atomic and molecular physics research in universities. Government agency and industrial research programs in this field were not included. The support of atomic and molecular physics research in government and industrial laboratories presents additional requirements beyond the scope of this recommendation.

Energy Commission, the Department of Defense, and other agencies that have traditionally supported this type of work.

RECOMMENDATION 3 Approximately $1.8 million, in addition to the level of support indicated in Recommendation 1, should be made available each year for the purchase of major capital equipment for atomic and molecular physics experiments.

RECOMMENDATION 4 The federal government should attempt to provide an additional annual increase in funds to (a) reduce the effects of inflation and (b) allow at least ten young scientists to enter atomic and molecular physics each year. This increase should amount to approximately $0.9 million per year, $500,000 to mitigate the effects of inflation and $400,000 for the support of new investigators.

RECOMMENDATION 5 The National Science Foundation should establish an effective method of supporting the research that is intermediate on the basic to applied continuum. Specifically, we *recommend* that the support of research in this intermediate area be closely coordinated with the efforts of the National Bureau of Standards, the National Institutes of Health, and other agencies having a strong interest in this type of research.

RECOMMENDATION 6 The National Science Foundation should give immediate attention to the effecting of greatly improved means of support for interdisciplinary basic research, apart from its several current programs in interdisciplinary applied research, particularly for large groups and major centers at universities, government laboratories, and industrial laboratories that are attempting to make a broad attack on complex research problems. The National Science Foundation also should make a special effort to ensure the continuity of established programs and interdisciplinary research groups that show promise of continued high productivity and aid in making these groups a resource for the scientific community in general.

2 Introduction

A. BACKGROUND OF THE REPORT

One purpose of this report is to acquaint both scientists and nonscientists with the content and achievements of atomic and molecular physics. Therefore, in Part II we describe the development of this field from the early discoveries in atomic spectra at the end of the last century to the many varied and complex experiments of today. In this review, we attempt to show how this field, so closely linked to the development of modern physics, relates to a variety of current practical and scientific problems that often have immediate bearing on our lives.

In addition, this report presents the results of a two-year study of the activities and current problems of atomic and molecular scientists. This study was begun in the spring of 1968, following increasingly frequent reports that abrupt terminations in federal support were halting the work of many of the most productive men in the field. Unlike most other physics subfields, atomic and molecular physics had at that time no spokesman—no committee charged with the responsibility to examine and alert the government to impending problems. If such problems actually existed, their dimensions were unknown. Consequently, it was essential to determine with some accuracy the size of the research effort in atomic and molecular physics, the degree of recent growth in experimental capability, the new ideas that merited exploration, the relationship of the field to major problems facing society and those studied in other disciplines, the magnitude and sources of research support, and the possible consequences of the reduction or termination of support for various programs.

With the approval of Donald F. Hornig, then the President's Special Assistant for Science and Technology, a nucleus of some ten atomic and molecular scientists designed a survey questionnaire and began to collect reliable information about the field. The Division of Electron and Atomic Physics of the American Physical Society encouraged and cooperated in this effort. Growing awareness of the study and its objectives stimulated a number of senior scientists, who also were concerned about the progress and future of this field, to volunteer their services. Soon some 50 of the most active atomic and molecular scientists in the United States were engaged in developing a detailed picture of the field and the problems it faces. As this report was nearing completion, the National Academy of Sciences–National Research Council established a committee to represent atomic and molecular physics. As a result, a mechanism for studying the activities and needs of atomic and molecular physics and fostering communication with the federal government was established. The findings of the Committee's survey and its recommendations appear in Part I.

This survey leaves many questions unanswered. However, when we consider that the communication of common problems was almost nonexistent among atomic and molecular physicists at the time that the survey began, it represents encouraging progress toward identifying problems of mutual concern and stimulating efforts toward their solution.

Much of the administrative information reported to the Committee cannot be accurately quantified because of incomplete records and differences in the way that the scope of atomic and molecular physics was defined. However, the qualitative picture is clear and accurate. It results from extensive interviews with scientists at virtually every university with research activities in this field and shows that atomic and molecular physics is expanding phenomenally in scope and opportunities for research and that the findings are of vital importance to progress in many fields of science. The results also show clearly that significant advances in many parts of the field are no longer possible because of lack of support. In addition, it is evident that federal research-support programs are far better geared to fields with relatively clearcut boundaries than to fields the major thrust of which is interdisciplinary and applied. Our recommendations address these problems.

B. ATOMIC AND MOLECULAR PHYSICS: ITS NATURE AND SIGNIFICANCE

What is atomic and molecular physics? And why should we be concerned about its welfare? The answers require careful study of the descriptions of the work that atomic and molecular scientists perform in the laboratory and are presented in detail in Part II. For the nonscientist, the field is perhaps best and most easily described by its uses and some simple examples.

The atomic and molecular physicist studies the forces between atoms to understand molecular structure and explain the nature of chemical reactions. Through ultraprecise measurements of certain atomic properties, for example, their energy, he can explore some of the ramifications of the structure of matter and even of "empty" space. In an extension of the techniques developed in the study of simple molecules, he is beginning now to study larger and more complex systems including the biologically significant "macromolecules." An understanding of the structure of these molecules is one of the basic goals of the life sciences.

Many atmospheric phenomena are directly attributable to atomic and molecular processes. For example, in regard to air pollution, we know that automobiles, power plants, and other combustion devices emit sizable amounts of sulfur, nitrogen, and organic compounds that take part in a wide variety of chemical processes with each other and with the normal constituent gases in the atmosphere. Eventually, these compounds are absorbed in the ecosphere so that the atmosphere somehow is cleansed after sufficient time has elapsed. Many atomic and molecular physicists are investigating the chemical history of atmospheric pollutants. With meteorologists, plant scientists, physical oceanographers, and other scientists, they are attempting to discover the way in which pollutant products are eventually absorbed.

Other atmospheric processes also require atomic and molecular physics for their explanation. The sun heats the earth's atmosphere and imparts thermal energy to its atoms and molecules. How solar energy is absorbed by these particles and transported from particle to particle through the atmosphere remains a mystery that challenges atomic and molecular scientists.

Many atomic and molecular physicists seek primarily an understanding of those basic phenomena that they find most interesting. However, they often discover that the knowledge gained in atomic and molecular physics is immediately applicable to broader problems in astrophysics, meteorology, biophysics, oceanography, and other disciplines and, as a result, can have a profound impact on problems affecting society. This interweaving of basic and applied science is a special characteristic of atomic and molecular physics; there is no clearcut boundary between pure and applied work for a great many atomic and molecular scientists.

On the other hand, not every important research area of basic atomic and molecular science has the same proximity to practical problems. For example, some work deals with the theory of relativity, which predicts that energy levels of even the simplest atoms are not what one would expect on the basis of nonrelativistic theory. The necessary corrections are extremely small and difficult to measure; they would seem to exemplify the kind of preoccupation with exceedingly fine points that can lead to meaningless obsessions. But these relativistic energy shifts have far-reaching implications throughout physics. Possibly, the theory of relativity will require major revisions that could have great long-range significance for more applied science. The only way to proceed is by testing every facet of its predictions.

This is but one of the kinds of basic experiment in which atomic and molecular scientists are engaged. Generally, such fundamental contributions require experimental methods of the greatest precision. Often atoms or molecules must be isolated from the perturbations of the outer world for substantial amounts of time. For example, in one experiment now in progress a technique had to be devised that would completely isolate a few atoms from all perturbations for days. Thus atomic physics, in what could be characterized as its purest form, continues to contribute vital information on the fundamental structure of matter.

Briefly, then, there are two extremes in atomic and molecular research activity, one characterized by the merging of pure and applied science and the other by a far-reaching impact on our understanding of the fundamental laws of nature. Between these extremes lie many types of studies. Part II provides a more balanced and detailed picture of what atomic and molecular physicists do.

3 Statistical Survey

A. SCOPE

The purpose of the statistical survey was to gather information on research and graduate training in universities. However, the unusually close interplay of university faculty and research programs with industrial and government laboratory personnel and research programs required some attempt to identify trends in these laboratories that might affect universities, job opportunities for new PhD's, and other related matters. A concerted effort to assemble such information met with little success; nor was even a rough indication of research, funding, and employment trends in atomic and molecular physics research in industrial or government laboratories possible. Atomic and molecular research rarely exists as an independent and permanent program in nonuniversity laboratories. In some cases, there may be a small nucleus of people who perform such research, but most of their activities usually are integrated into broader functional groups having applied research objectives. Management is rarely interested in, or aware of, the division of effort between atomic and molecular research on the one hand and, for example, solid-state research on the other. The resulting lack of documentation makes it virtually impossible at this time to delineate those industrial and governmental activities that could be considered comparable to university atomic and molecular research.

A variety of unexpected problems occurred during collection of the university statistics. Some difficulty resulted because atomic and molecular research is not the exclusive province of any single academic department. It is actively pursued in physics, chemistry, biology, engineering, and other departments, each discipline having its particular goals and research styles. Discussions with university faculties from a variety of disciplines, and with many officials at federal agencies, verified that the problems and support trends experienced in physics departments are characteristic of atomic and molecular research in all disciplines. After careful review of the initial responses, the Committee decided to limit the scope of this survey to physics departments to avoid serious ambiguities. However, the Committee recognized that this decision would necessarily exclude much excellent and significant atomic and molecular research in other groups, particularly within chemistry departments. Any future studies undertaken in this field should attempt to correct this narrow focus by including relevant activities in other university groups.

B. DEFINITION AND PROCEDURE

A working definition of atomic and molecular physics is difficult to achieve. The dividing line between subfields within physics is indistinct, and the boundaries in chemistry are even more difficult to delineate. One approach is to assemble a list of topics that presumably constitute atomic and molecular research, but such lists become so technical and require so much interpretation and analysis that they generally create as many questions as they answer. In preference to this procedure, the survey committee selected a senior representative for each campus who would resolve questions relating to the definition of the field on the basis of his judgment and ensure that the data for his institution were inclusive. In addition, the local representative could identify important factors and trends affecting a university. Adoption of this procedure eliminated the necessity for a highly structured questionnaire in which conceptual errors of the survey committee might seriously influence the results. Although the replies from various institutions would differ in certain respects as a result of this method, and representatives might fail to identify all the important problems in their institutions, the procedure offered the advantage of thorough and intelligent consideration by local representatives as opposed to undue concern with lists of definitions and rules.

A local representative was selected for each of the 104 institutions known to be active in atomic and molecular research. This list of institutions was checked against a recent survey of fields of graduate research offered at PhD-granting institutions compiled by the American Institute of Physics.[1] Each representative received a copy of the questionnaire that appears in Appendix I.A. This questionnaire was designed to induce respondents to look beyond the specific questions it included. Individual questions were worded in such a way that they required some interpretation and thought by the reader. No specific data form was provided, thus each representative had to consider whether the inclusion of additional data would be useful and meaningful. (This procedure was not entirely successful, as few undertook the additional effort required for the inclusion of more comprehensive data.)

The first survey questionnaires requesting information on the academic year, 1967–1968, were mailed to 92 institutions in June 1968. By the end of August 1968, 46 replies had been received, of which 32 were sufficiently complete to be used. The other 14 were incomplete because of the difficulty of extracting reliable information, a difficulty that resulted from (a) temporary lapses in research support, (b) the lack of separate records for each investigator in cases of institutional or group support, and (c) the practice of conducting research in several fields or subject areas with support from a single grant.

These problems were discouraging but not insurmountable. The Committee found that fairly reliable data could be obtained by encouraging investigators with complicated support structures to estimate their expenditures. This procedure seemed justifiable since the greatest record-keeping problems existed on campuses having the largest volume of research support and the highest levels of research activity. In these cases, we assumed that individual errors of estimate would compensate one another statistically when the data were summed.

A second attempt to complete the survey took place early in 1969 using the revised procedure. Rather than making another attempt to obtain information from undergraduate institutions, the goal of this

[1]*Summary Publication* R-205, which accompanies AIP Publication R-205, *Graduate Programs in Physics and Astronomy* (American Institute of Physics, New York, 1968).

second effort was to achieve complete coverage of all major graduate institutions.

C. RESULTS AND INTERPRETATION

Replies were obtained from 55 institutions, 53 of which offer graduate training. The results appear in Table 1. Though replies were received from every major graduate center with activities in atomic and molecular physics, only 194 senior investigators were reported, which appeared to be too small a number in comparison with the approximately 350 faculty members reported by the Panel on Atomic and Molecular Physics and Quantum Electronics of the Physics Survey Committee in their 1966 report.[2] However, the Physics Survey Committee used methods that would produce an overestimate of the number of faculty members who were active in research and a corresponding underestimate of the average support levels for graduate research programs. In the survey reported in 1966, questionnaires were sent to all university and college faculty members known to have been active in atomic and molecular research at some time in their careers. The data relating to physics departments showed that 57 investigators, of a total of 247, were listed as active but had no support. Among them were several very eminent scientists who command a great deal of support. Six other investigators, who together in 1965 received $391,000 of the total of $9.122 million, were included as a result of obvious misinterpretations of the questionnaire. (Three were plasma physicists, one was a nuclear physicist, one was a many-body theorist, and one, a solid-state physicist.)

The Physics Survey Committee figures for nonphysics university departments, government laboratories, and industrial laboratories contain similar ambiguities and raise doubts in regard to the conclusion that the average annual support in all types of laboratory per principal investigator was $39,400 in 1964. A careful study of the Physics Survey data suggests that the average annual support per active principal investigator engaged in atomic and molecular research in the physical sciences in 1964 was at least $47,500.

Average support figures of this sort often have little meaning until the information has been broken down by class of institution so that

[2]Physics Survey Committee, NAS–NRC, *Physics: Survey and Outlook*, NAS–NRC Publ. 1295 and 1295A (National Academy of Sciences–National Research Council, Washington, D.C., 1966).

TABLE 1 Statistical Summary–University Atomic and Molecular Physics

Categories of Data Collected	1964 Physics Survey[a]	1968 Committee[b] Survey	Additional 1968 Government Data	Total 1968
Total university physics support of respondents ($ millions)	$ 8.731	$ 9.017	$ 1.864	$ 10.881
Number of senior investigators reported	184	223	79	302
Average annual support/ senior investigator ($ thousands)	$ 47.5	$ 40.5	$23.6	$ 36
Number of postdoctoral students of respondents	160	137	–	–
Average postdoctorals/senior investigator	0.87	0.61	–	–
Number of students/senior investigator	–	2.6[c]	–	–

[a]Adjusted as specified in the text in Chapter 3, Section C.
[b]The Committee on Atomic and Molecular Physics of the Division of Physical Sciences of the National Academy of Sciences–National Research Council.
[c]Of these, 393 were supported by grants and contracts (68%), 105 by fellowships (18%), and 76 by teaching assistantships (13%).

the reader can compare either the situation at the largest graduate institutions with that at smaller graduate schools or the cost of research in a graduate institution with that in an undergraduate college. An examination of the conduct of atomic and molecular research in the smaller graduate schools indicates that an average program at a major center commands about three times the support normally needed per program at smaller, less-developed graduate institutions. The primary difference is that the most modern types of research demand the continuing acquisition of highly sophisticated optical and electronic equipment and the fabrication of a great deal of additional equipment not commercially available. Although modest programs with very little new equipment continue at smaller institutions, such programs generally are approaching scientific obsolescence. The investigators involved are often trapped in the predicament of lacking sufficient new equipment to perform the feasibility studies needed to seek support for more modern programs. The 3 : 1 ratio actually represents a vast difference in the quality and innovative nature of research; the research careers of those receiving low support are slowly stagnating. Most scientists involved in such programs are aware of the difficulties and frustrations of their position but prefer doing at least some research to doing none.

Undergraduate research programs also generally are near, if not beyond, the point of obsolescence in terms of innovative approaches. They differ in cost from the least expensive graduate programs largely in the lack of graduate student and postdoctoral stipends. Nevertheless, in some cases there is little substantive difference between undergraduate and the less-expensive graduate atomic and molecular research. A few striking examples can be found in which undergraduate research represents the highest standards of quality in the field; however, most of the investigations performed under these funding limitations are, of necessity, very modest in scientific yield.

Small-scale programs can be quite worthwhile if care is exercised in limiting study to standard methods of examination of systems in which more comprehensive data are needed. Because the increasing need of industry and science for accurate atomic and molecular data generally can be met only by making direct measurements, such small-scale programs often produce highly useful and valuable data. Few of the recent major advances in atomic and molecular physics could have been achieved without accurate data on atomic and molecular systems. One example is the invention of the maser, which resulted directly from comprehensive measurements of energy levels of the ammonia molecule taken at the highest possible resolution.

Unfortunately, increases in technological sophistication generally have caused most small-scale, low-budget programs to become outmoded, especially those in undergraduate institutions. This trend contrasts strongly with the situation that formerly existed, for atomic and molecular research only recently has begun to require large support. Undergraduate institutions often made important contributions to the field. The present survey did not attempt to obtain undergraduate data except from a few institutions in which high-quality research is maintained. Their replies, in addition to informal inquiries, indicated that the inclusion of more data from undergraduate institutions would produce results that did not represent the problems affecting investigators attempting to maintain programs appropriate to the graduate level.

Viewed in this context, this survey's inclusion of only about 60 percent as many atomic and molecular scientists as were reported in the 1966 Physics Survey Committee report—although five years elapsed between these surveys in which substantial population growth might have occurred—does not significantly affect the findings. Most, if not all, graduate institutions emphasizing atomic and molecular physics research are included. Therefore, the results should provide a highly representative picture of graduate research conditions in physics departments in 1968. (Data on chemistry departments were not sufficiently comprehensive to be included in the current report.)

In principle, these 1968 physics department figures, showing 223 principal investigators with an average annual support of $40,500, should not be compared with the 1964 Physics Survey data. This later survey was directed primarily toward the 53 largest graduate institutions and did not attempt to include those departments having only small or token programs, whereas the Physics Survey included all sizes of institutions.

The Committee supplemented its survey data through the inclusion of some unofficial data available from federal agencies.[3] These data extended coverage to an additional 45 graduate and undergraduate institutions, bringing the total to 100. The total for the faculty increased from 223 to 302. As might be expected, the additional 79 investigators generally commanded less support than their colleagues at large institutions, the average for these 79 being $23,600 per investigator per year. When the federal agency data for the additional 45 institutions are added to the survey data, an average support figure of $36,000 per principal investigator per year results. This is probably a reasonable estimate of the over-all average support for 1968 and should be comparable (in regard to the size and composition of the population on which the calculation is based) with the adjusted 1964 Physics Survey Committee annual support figure of $47,500 per principal investigator in physics departments. Although the method of calculation is not exact in either case, apparently a reduction in annual support of about $10,000 per atomic and molecular physics investigator occurred between 1964 and 1968.

The cost of atomic and molecular physics research in graduate institutions has increased markedly in recent years. A minimum program staffed by one professor and two students now costs about $40,000 per year, whereas a program of average size costs about $76,000 per year. (See Tables 2 and 3 for sample budgets.) The cost

[3]*Report #2 of the Interagency Group of Atomic, Molecular and Plasma Physics Administrators* (an informal group formed by L. T. Crane and including representatives of the U.S. Army, Navy, and Air Force, the Atomic Energy Commission, and the National Science Foundation).

TABLE 2 Sample Budget (One Year) for a Small-Scale Experimental Atomic and Molecular Program[a]

1. Senior salaries			$ 3,200
A. Principal investigator (2 months, summer)		$3,200	
2. Other salaries			10,625
A. Graduate stipends (2 @ $4,000)		8,000	
B. Technicians and shop (3 man-months)		2,025	
C. Secretarial (10% time)		600	
3. Overhead (50% of salaries)			6,913
4. Salaries and overhead	$20,738		
5. Expendable equipment			7,000
6. Permanent equipment[b]			10,000
7. Computer			1,000
8. Miscellaneous			1,500
TOTAL			$40,238

[a]This sample budget is not intended to serve as a model grant budget. It includes all expenses to the research project, including some that are normally contributed by the host institution.
[b]This term is explained in the text (Chapter 3, Section C). Briefly summarized, this budget item provides $5,000 per year for relatively small pieces of standard laboratory equipment and a $5,000 allowance for the fact that a major new piece of equipment (costing about $30,000) generally is required every third year.

TABLE 3 Sample Budget (One Year) for an Average-Sized Experimental Atomic and Molecular Program[a]

1. Senior salaries			$ 3,800
A. Principal investigator (2 months, summer)		$ 3,200	
2. Other salaries			21,600
A. Postdoctorals (1)		10,000	
B. Graduate stipends (2 @ $4,000)		8,000	
C. Technicians and shop (4 man-months)		2,700	
D. Secretarial (15% time)		900	
3. Overhead (50% of salaries)			12,700
4. Salaries and overhead	$38,100		
5. Expendable equipment ($7,000 per student)			14,000
6. Permanent equipment[b]			20,000
7. Computer			2,000
8. Miscellaneous			2,000
TOTAL			$76,100

[a]This sample budget is not intended to serve as a model grant budget. It includes all expenses to the research project, including some that are normally contributed by the host institution.
[b]This term is explained in the text (Chapter 3, Section C). Briefly summarized, this budget item provides $10,000 per year for relatively small pieces of standard laboratory equipment, and a $10,000 allowance for the fact that a major new piece of equipment (costing about $30,000) generally is required every third year.

in 1964 was considerably less. Although salaries have risen in concert with other aspects of the economy, this item is only a minor cause of growing costs. The major cause is the need for the highly sophisticated approach and instrumentation required in modern research. Advanced data-handling systems, tunable laser sources, mass spectrometers, spectrophotometers, and other such items are essential to atomic and molecular research; many of these instruments cost tens of thousands of dollars. An investigator contemplating an apparently small extension in his research often is faced with the need for additional equipment amounting to about $30,000. This situation occurs even in laboratories in which some equipment can be borrowed from other investigators. Thus the sample budgets appearing in Tables 2 and 3 include $5,000 and $10,000, respectively, per year for large equipment items on the assumption that at least one such instrument will be needed every third year.

Larger programs, of course, require substantially greater support. The upper limit of the range in 1968 was $200,000 for a single investigator, two postdoctoral fellows, and six graduate students working in quantum electronics. One group of five senior investigators working together in the same field required $500,000 per year. However, a typical atomic and molecular physics program now requires about $90,000 per year to run efficiently with a staff of one senior investigator, one postdoctoral student, and three graduate students.

Computer technology has created new opportunities for atomic and molecular research; it also is likely to lead to a substantial escalation in the level of support required to conduct reasonably up-to-date research. Large amounts of computer time are essential for theoretical analysis and data reduction, and small computers are needed to provide on-line control of many difficult experiments that could hardly be performed without computer assistance. A similar revolution in methodology, of course, has occurred also in nuclear and high-energy physics, but the large ongoing cost of experiments in these fields conceals, in large measure, the magnitude of the expense entailed in using computer technology.

Atomic and molecular theoretical scientists are examing many major problems that require large amounts of computer time. The subtle energy balances that determine the chemical processes of life, the behavior of much of the atmosphere and outer space, and the structure of atoms require the extensive use of computers. Increasingly often we find that an atomic and molecular theorist needs 25 hours of computer time per year, which constitutes an annual cost of about $10,000.

4 Conclusions

A. BACKGROUND

It would be difficult to argue that the value of atomic and molecular physics lies in the promise of some undiscovered physical law that might challenge all of physics as did the quantum theory. Such a discovery could occur, but it appears unlikely at this time. Much the same is true of most other physics subfields. The frontier on the totally unknown lies predominantly in elementary-particle physics; there radically new findings can be expected. However, among the physics subfields having a clear and demonstrable relationship to problems of immediate national concern, atomic and molecular physics is pre-eminent. This subfield is related to the chemistry of life processes, the mechanisms of disease, the processes underlying all types of pollution, the fluid dynamics of the atmosphere and oceans, and many problems ranging from power production to missile reentry in the atmosphere. The solution of these problems will require a vast amount of work in atomic and molecular science. Outstanding experimental and theoretical problems demand the best talent that the community of basic research scientists can provide, although the fundamental physical laws presumably are well understood.

In the past physicists directed most of their attention to the achievement of an understanding of the interactions among a few particles or the behavior of simple systems. Current challenging problems demand knowledge of complicated systems, such as many-electron atoms and large ensembles of atoms, ions, or molecules. The solutions lie not in reorganizing and recombining existing physical knowledge but in seeking an understanding of situations in which varieties of processes compete with one another to produce macroscopic events, which is the province of atomic and molecular physics and the physics of fluids.

In a scientific community that values the traditions of elegant simplicity in theory and experiment, many of the problems of greatest interest to atomic and molecular scientists appear inelegant. The so-called many-body problems, which characterize so much of current atomic and molecular research, are not amenable to solution with the precision customarily sought in basic sciences. This is one of the distinguishing features of atomic and molecular research. Because atomic and molecular scientists are challenged increasingly by the physics underlying practical problems, much of their work is not generally regarded as basic research.

On the other hand, scientists performing highly applied research often feel that atomic and molecular scientists are directing their attention to matters that are too fundamental to be immediately applied. This view is unfortunate and endangers atomic and molecular research, which has major implications for molecular biology, pollution problems, weather prediction, and a host of other matters of vital significance to society. Yet, in an era that hails the importance of relevant science, atomic and molecular scientists find it difficult to obtain support for their work.

The data on funding trends in atomic and molecular research collected in the Committee's survey define this problem. The average support per active principal investigator in atomic and molecular physics declined by about $10,000 between 1964 and 1968, or by about 6.5 percent per year. At the same time, the rapid development of sophisticated new equipment together with increasing costs of salaries and expendable items raised the minimum annual cost of a modest program by $10,000 to $15,000. This finding indicates that those investigators who still receive support have experienced an effective reduction in level of support of 12 percent to 15 percent per year during the interval from 1964 to 1968. These trends in costs and funding occurred during a time in which atomic and molecular scientists were beginning to make progress toward understanding the physics that underlies the solution of many major national problems.

Some atomic and molecular investigators lost all support during this same period. Not all of these terminations resulted from poor-quality research. Atomic and molecular spectroscopy, so vital to device development, atmospheric measurements, and many defense-related problems, suffered badly.

Although reductions in federal support are widespread in the scientific community, atomic and molecular scientists have experienced particular difficulty in obtaining support. No funding agency gives specific attention to maintaining and promoting this field as basic research, a situation in marked contrast with that characterizing solid-state physics, nuclear-structure physics, elementary-particle physics, and the recently emerging subfield of intermediate-energy physics, for which support programs exist at the AEC, Advanced Research Projects Agency (ARPA), NSF, and, until recently, the Navy. The NSF is the only agency with a program that specifically includes atomic and molecular physics. Formerly, this program was designated Atomic and Molecular and Other Physics,* thus signifying its responsibility for all experimental physics not included in solid-state, nuclear-structure, intermediate-energy, or high-energy physics. It continues to include—in addition to atomic and molecular physics—plasma physics, quantum optics, chemical physics, acoustics, optics, and experimental investigations of basic natural laws in relativity and quantum electrodynamics. The Program attempts to meet these responsibilities with less than 10 percent of the NSF Physics Section budget for research and facilities support, a fraction that has been roughly constant since fiscal year 1963.

The U.S. Air Force, Navy, and Army also have supported fundamental atomic and molecular research, primarily through small programs with responsibilities as diverse as those of the NSF Atomic, Molecular and Plasma Physics Program. Physicists at federal agencies often have attempted to relegate major portions of atomic and molecular physics research to those offices responsible for chemistry, astronomy, biology, atmospheric sciences, or other developmental efforts. Their opinion was that the users of atomic and molecular data should provide substantial support for this field, which still poses some of the most challenging problems in fundamental physics.

Atomic and molecular research survived and grew for the reasons that appeared to be its greatest liabilities: It was so essential to health

*Now titled the Atomic, Molecular and Plasma Physics Program.

and defense interests, and subsequently to NASA's interests, that atomic and molecular scientists in both industry and universities were necessary to develop the basic research knowledge required in the solution of many highly applied problems. Such applied needs provided a great deal of the motivation for the dramatic growth of atomic and molecular physics as a basic as well as an applied science in the last 20 years. (A striking number of prominent atomic and molecular physicists now working in universities made their initial reputations in industrial or government laboratories.)

Some funds (about $600,000 annually) for the support of atomic and molecular research came from the Joint Services Program. In the early 1960's, NASA attempted to provide support for all space-related science through a myriad of objective-related offices and a Sustaining University Program. This Program, since terminated, reached a peak level of $46 million a year. The Air Force's Cambridge Research Laboratory, Wright-Patterson Air Force Base, and many of its other scattered development centers each supported a small amount of research related to their missions. For many years, a similar situation existed in the Navy and, on a lesser scale, in the Army. The Advanced Research Projects Agency indicated a broad interest in atomic and molecular research and established a number of university-related institutes to promote its interests. This agency also supported a large number of individual projects. The combined support of these various agencies was sufficient to maintain a healthy amount of basic atomic and molecular research from the early 1950's through the early 1960's.

Other physics subfields obtained support from many of the same sources during that period, but none was forced into such a heavy reliance on offices directly responsible for developmental programs and none suffered the consequences of funding cuts as severely. Early in the 1960's, it was apparent that the immediate research needs of current defense and space programs would decrease for the next several years. Budget reductions in all federal offices demanded curtailment of support for activities not related to immediate missions. Offices responsible for developmental research concluded that most basic research in universities was expendable. Often a 5 percent or 10 percent reduction in the budget of such an office would precipitate a 100 percent reduction in basic research support. The pressure was increased by growth of the philosophy that basic research support should be the responsibility of the NSF rather than being included in developmental efforts. Figure 1 depicts the generally declining levels of support for atomic and molecular physics and plasma physics during the late 1960's.

It is impossible to assess the exact amount of atomic and molecular research that has been curtailed. There was little awareness in the government of the effects of these budgetary and policy pressures on basic research, except in such cases as the Navy's transfer of its nuclear-physics program to the NSF and the impending transfer of the ARPA Materials Research Program and the Air Force National

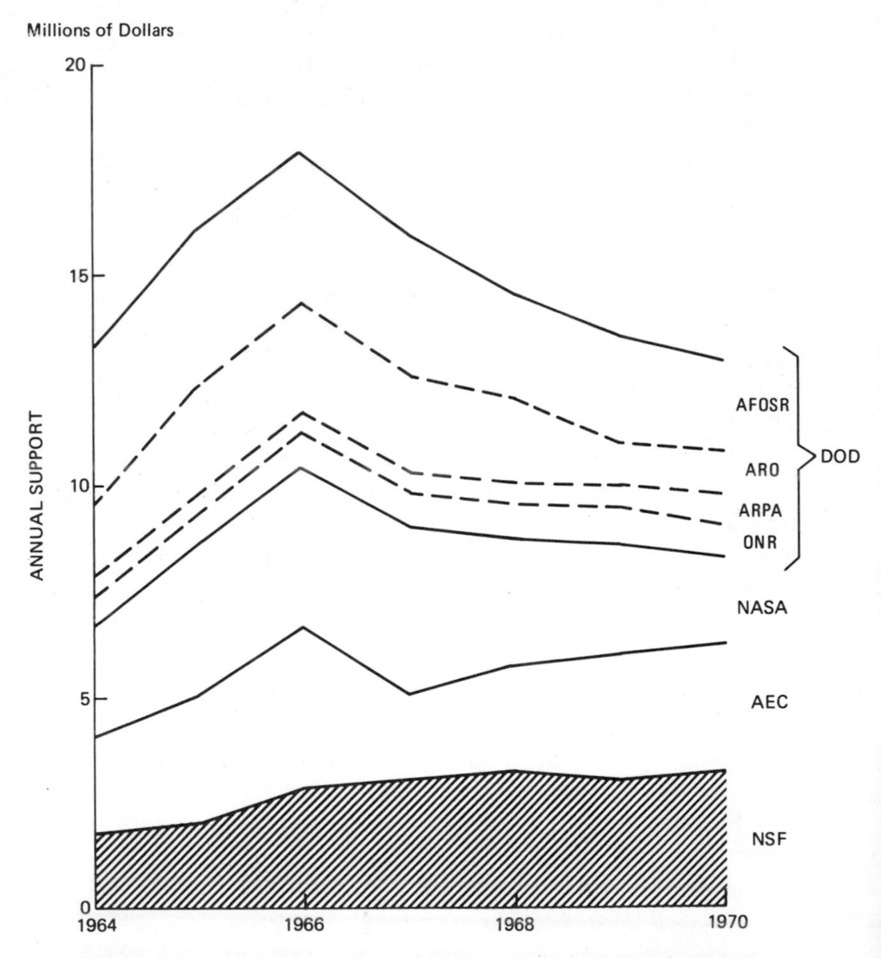

FIGURE 1 Federal support of university research in atomic, molecular, and plasma physics. (These statistics were collected informally from federal agencies by the Physics Section of the National Science Foundation.)

Library
I.U.P.
Indiana, Pa.
537.1 N2134a
C.1

Magnet Laboratory to the NSF, all of which have been carefully planned and negotiated at the highest levels. To date, the NSF has not sought a budget increase to cover known terminations of support by other agencies for atomic and molecular physics research. At the end of 1968, a fairly large number of atomic and molecular investigators were working under no cost extensions of rather old contracts by carefully conserving funds so that students could complete their research. Many of these investigators were without funds by 1970. Perhaps most serious from the standpoint of government objectives was the forced curtailment of those atomic and molecular programs in which basic and applied scientists were engaged in the type of cooperative efforts that the government has attempted to stimulate during the past 20 years. Work on applied problems, which long provided intellectual stimulation and financial support for many atomic and molecular scientists, resulted in funding problems for some of the most outstanding scientists in this subfield in the late 1960's.

B. RECOMMENDATIONS

In spite of its relevance to a great many national problems, from the SST to pollution to molecular studies of life processes, it is becoming increasingly difficult for universities to train students in atomic and molecular physics. The disappearance of student stipends and research support creates severe difficulties in almost all scientific disciplines; however, the effects are much more severe in atomic and molecular science because of several unusual factors discussed in the preceding section. A major one is the former heavy dependence of atomic and molecular research on many relatively small mission-oriented federal offices for support. The termination of support for basic research relating to project missions ended the work of many investigators, some of whom were widely known for their outstanding research. A second major factor was the rapid growth of interest in atomic and molecular science in the last half of the 1960's. New exceedingly high-precision experimental techniques opened vast new areas of investigation with great theoretical and practical potential. For example, the laser provided challenging new optical sources, and new scattering techniques allowed scientists to examine the subtle dynamics of chemical processes in action. Many similarly spectacular breakthroughs in experimental methodology gave added meaning and importance to the field.

As a result, after a long period of relatively low activity in atomic and molecular physics research in universities, it suddenly became the focus of substantial interest. Universities at which groups of atomic and molecular scientists worked found it necessary to initiate some of the newer types of experiments if they were to remain current in this field. Many developing universities either began atomic and molecular research projects or greatly expanded the scope of their efforts.* Thus atomic and molecular research expanded rapidly and was not so dominated by large institutions as to preclude meaningful contributions from newer and smaller universities. Currently, a number of new experiments, particularly at recently developed physics and chemistry departments, are partially built or under way but lack permanent support; they represent a substantial resource that could be lost to the nation unless funding assistance is soon available.

Although it is difficult to obtain a reliable figure for the total support needed, a minimum estimate results from an analysis of recent actions of the Atomic, Molecular and Plasma Physics Program of the NSF.† This Program is responsible only for physicists; it does not include atomic and molecular scientists working in such disciplines as chemistry, biology, meteorology, engineering, and the like. It funds only experimental research; theoretical work in atomic and molecular physics is supported by another NSF program. The 1971 budget of the Atomic, Molecular and Plasma Physics Program is about $2.8 million, which provides funds for 87 projects at an average annual rate of about $32,000. This amount is approximately 50 percent of the minimum level cited in the previous section as being required for the smallest atomic and molecular physics programs. (The situation for atomic and molecular physics is actually somewhat worse than the figure of $32,000 implies. Approximately 25 percent of the grants of the Atomic, Molecular and Plasma Physics Program are for the support of plasma physics; these projects require and receive a higher average level of support than those in atomic and molecular physics.)

*For example, University of Arizona, University of California at Santa Barbara, City University of New York, Clarkson College of Technology, University of Colorado, University of Maryland, University of Massachusetts, University of Nebraska, University of New Hampshire, New Mexico State University, University of Pittsburgh, Rice University, University of Texas, University of Toledo, and Wayne State University.

†We have not attempted to separate the plasma physics support, since the NSF currently supports atomic and molecular physics and plasma physics through a single section.

The demand for support in atomic and molecular physics research greatly exceeds the resources of the NSF Program. After applying the most stringent quality standards to incoming proposals, the Program lacked funds to support 126 separate proposals amounting to an aggregate annual level of support of $5.2 million for the combined two-year period, fiscal years 1970 and 1971. The average annual support sought in these proposals was small—only approximately $41,000; they represented desperate pleas of scientists for sufficient funds to remain professionally active, not to establish exciting new programs. New research programs would have required average support levels of at least $60,000 per year, or 50 percent greater support than the amounts requested. The $20,000 difference reflects the virtual impossibility of obtaining funds for capital equipment; therefore, investigators are limiting their efforts to experiments that can be conducted with equipment that is obsolete by most modern standards. It is incorrect to assume that this practice will merely prolong the time required to complete the experiments, for modern research can rarely be done with outmoded equipment. Thus atomic and molecular physicists have ceased trying to do the most up-to-date and potentially useful work and are merely trying to remain professionally active.

All support programs at the NSF currently are deluged with proposals, but few can match the excessive demand for support experienced by the Atomic and Molecular Physics Program. Until fiscal year 1967, this Program was devoted largely to helping small, preferably undergraduate, institutions to develop some type of research activity. While most other research support programs of the NSF were attempting to stimulate the fields for which they were responsible to the highest levels of research attainment, the Atomic and Molecular Physics Program (as it was then called) was preoccupied with the problems of establishing and improving science education at relatively weak institutions. As a result, the NSF budget for atomic and molecular physics did not grow, nor did it reflect the needs of the field.

In fiscal year 1967, the Atomic and Molecular Physics Program changed its direction and began to operate under the same principles as did other programs. However, the opportunity for significantly increased support of atomic and molecular physics by the NSF had already passed. All federal agencies, including the NSF, began to reduce research support budgets in fiscal years 1967 and 1968. Although atomic and molecular scientists turned increasingly to the NSF for support as a result of the changed emphasis of the Program, some two years elapsed before this agency fully realized that its support of atomic and molecular physics was inadequate. By that time, the problem facing the Atomic and Molecular Physics Program had become overwhelming. Not only was atomic and molecular physics in a stage of unprecedented growth and intellectual excitement, but the relatively new and expensive field of plasma physics suddenly emerged as a major new research activity on university campuses. There was no way to meet this combined problem without a major infusion of funds.

The present combined support problem of atomic and molecular physics and plasma physics far exceeds the budget of the NSF Atomic, Molecular and Plasma Physics Program. Since the middle of fiscal year 1968, the Program has not been able to initiate support of a new project without terminating that of another good program. Figures 2 and 3 illustrate the problem and show the number and quality of proposals supported and declined by the Program in 1967 (just after the Program began to foster the best research) and 1970. In fiscal year 1970, with a budget of about $2.5 million, the Program provided support for *only one new proposal.* It had to decline 69 high-quality proposals that merited support at an aggregate annual rate of about $2.7 million. In 1971, the Program expects to decline at least 45 new proposals that merit support and constitute a combined request of more than $2.5 million per year. (These figures have been corrected to exclude those scientists whose proposals were declined in 1970 and resubmitted in 1971.)

An examination of the proposals that were not supported shows that there are at least 126 experimental scientists in universities who are attempting to work primarily in atomic and molecular physics and related areas and who need support amounting to a minimum of $5.2 million annually to conduct their research. An additional $1.5 million is necessary to upgrade the support of atomic and molecular experiments already funded by the NSF. These data do not include the needs of atomic and molecular theoretical physicists, whose proposals are handled through another program in the NSF Physics Section. These theoretical atomic and molecular physicists need an additional $1.5 million per year to continue to be active in research. Therefore, if university-based atomic and molecular physics is to be maintained in a healthy state and is to be capable of contributing to national goals, approximately $8.2 million in additional funds is needed each year. Present levels of support will lead only to stagnation of research in this field.

FY 1967

NEW RESEARCH PROPOSALS–DECLINED

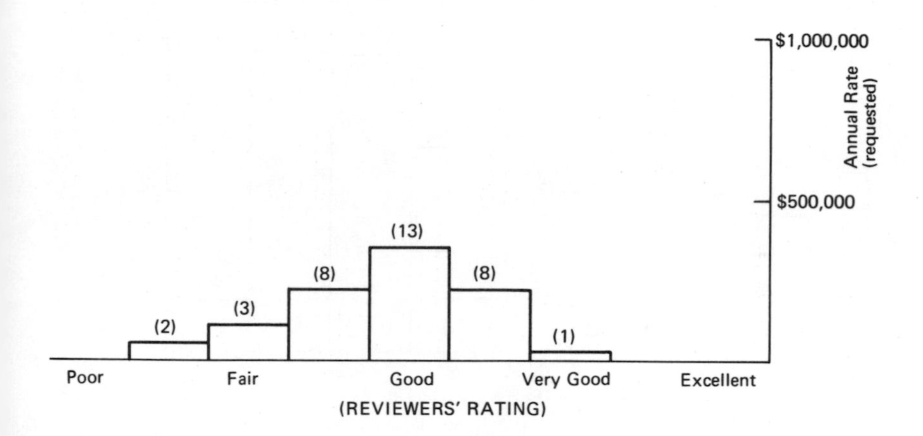

FY 1967

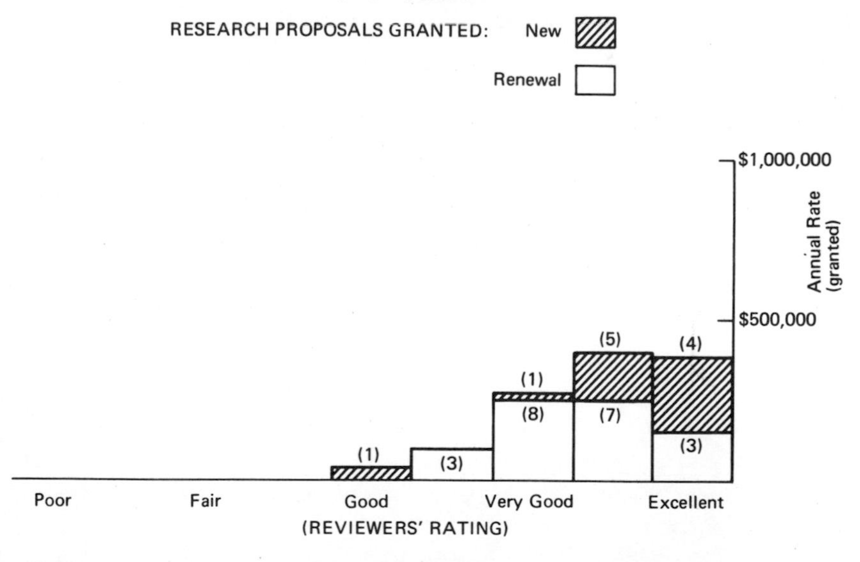

FIGURE 2 New research proposals declined and research proposals granted by the National Science Foundation in atomic, molecular, and plasma physics in 1967. (The "Reviewers' Rating" is a simple average of their opinions. It is not the only selection criterion and is used here only to generate an abscissa.)

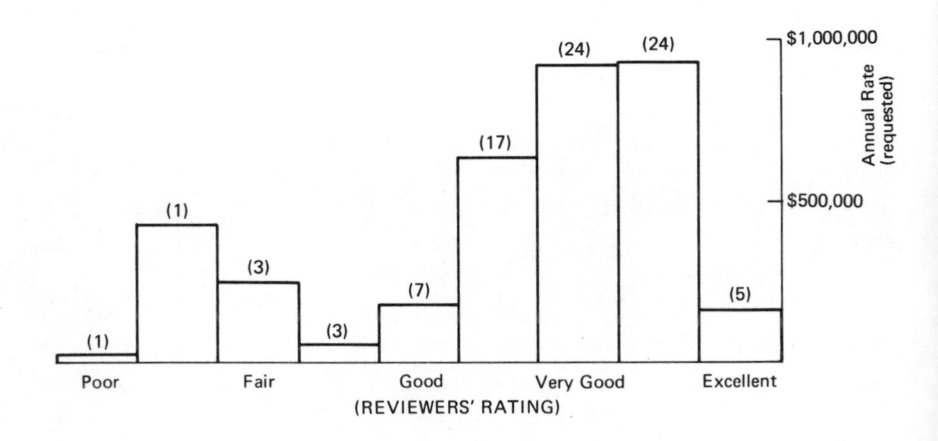

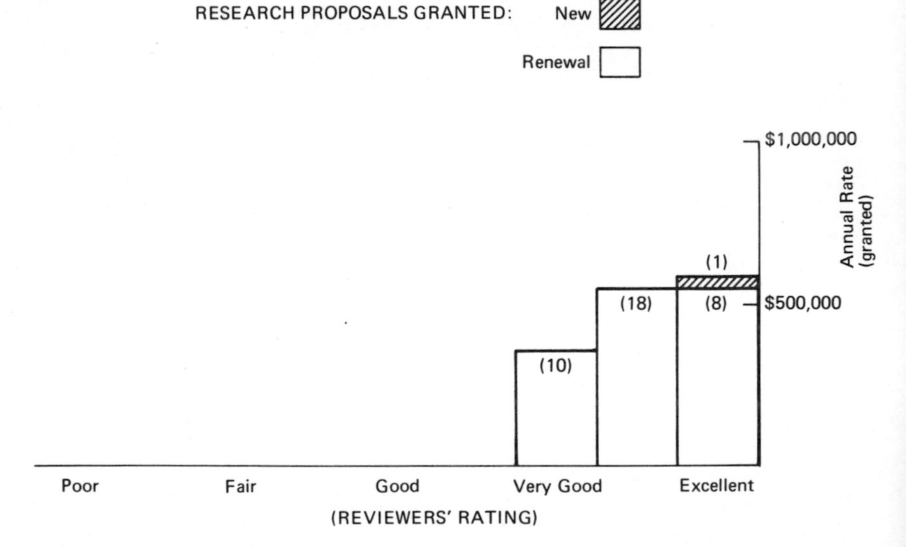

FIGURE 3 New research proposals declined and research proposals granted by the National Science Foundation in atomic, molecular, and plasma physics in 1970. (The "Reviewers' Rating" is a simple average of their opinions. It is not the only selection criterion and is used here only to generate an abscissa.)

The budget proposed for the NSF Physics Section in fiscal 1972 exemplifies the continuing low priority accorded to the needs of atomic and molecular physics. As Table 4 shows, the Atomic, Molecular and Plasma Physics Program received the smallest increment by far of all the Programs in the Physics Section; it continues to command only 8.6 percent of the total NSF funds devoted to the support of basic physics research. Although all physics subfields currently are having to cope with the effects of budget cutbacks in a number of agencies and to meet present demands for relevance in research, it is especially important to develop a healthy level of effort in subfields such as atomic and molecular physics that will play a major part in the solution of current national problems. The NSF should provide a substantial increase in its support of atomic and molecular physics, and other mission-oriented agencies should restore their far-sighted support of basic atomic and molecular physics as a means of ensuring that the goals of their missions will be met. Therefore, the Committee recommends the following:

RECOMMENDATION 1 Federal support of basic atomic and molecular research in universities should be increased by at least $8.2 million in the next three to five years to ensure that current levels of activity will not be reduced.

In the future, applied technological needs will continue to require the maintenance of a reservoir of new basic knowledge in the specific areas on which particular technologies draw heavily. An agency

TABLE 4 National Science Foundation Physics Section Budget

Program	FY 1971 ($ millions)	FY 1972 ($ millions)	Increase ($ millions)
Atomic, molecular, and plasma physics	2.7	3.6	0.9
Elementary-particle physics	10.8	14.7	3.9
National Magnet Laboratory	–	–	–
Nuclear physics	9.0	11.3	2.3
Solid-state and low-temperature physics	4.7	6.0	1.3
Theoretical physics	3.8	5.6	1.8
TOTAL	31.0	41.2	10.2

whose primary charge is to support basic science as such is not always in a position to allocate its resources in such a way as to guarantee the optimum rate of growth (or sustained productivity) in specific fields that support the specialized needs of other agencies. Therefore, to maintain the necessary production of new information and a reservoir of active, enthusiastic, and capable manpower in the fields on which they rely, many of the government agencies concerned with applied problems should give direct support to the continued growth of basic knowledge in such fields. In the past four years, budgetary constraints, often coupled with too narrowly construed requirements of relevance, have reduced some of these essential programs of support of basic research to dangerously low levels. Unless such basic-research support programs are revived promptly, a number of agencies could soon face a severe shortage of certain types of information and scientific expertise as they attempt to meet urgent and critical new problems.

RECOMMENDATION 2 Atomic and molecular science has received significant stimulation from its close association with and support through applied research programs in the federal government. Current support reductions in many of these programs have jeopardized their effectiveness and that of atomic and molecular physics research. Therefore, the federal government should make a concerted attempt to revitalize the support of basic research through applied programs in the National Aeronautics and Space Administration, the Atomic Energy Commission, the Department of Defense, and other agencies that have traditionally supported this type of work.

In addition, there is urgent need for approximately \$1.8 million annually for the purchase of large capital items. Atomic and molecular physics is experiencing such rapid growth that the obsolescence rate for equipment currently is extremely high. If growth patterns in other fields are indicative, this trend toward rapid obsolescence of equipment can be expected to continue for at least the next ten years.

RECOMMENDATION 3 Approximately \$1.8 million, in addition to the level of support indicated in Recommendation 1, should be made available each year for the purchase of major capital equipment for atomic and molecular physics experiments.

Atomic and molecular physics also must be able to expand. Few young people entered this field between about 1945 and 1960, when general interest in it was low. Now that many of the older atomic and molecular physicists have retired or died, most scientists in this field are below 45 years of age. Atomic and molecular physics requires a continuous influx of young scientists if it is to experience steady progress and avoid repeated cycles of activity resulting from uneven age distributions. If federal support is fixed at current levels, the unusually low death and retirement rates occurring as a result of the present age distribution will make it almost impossible for young scientists to enter this field. This problem recently was exacerbated by the termination of a number of atomic and molecular programs in industry; as a result, some excellent senior scientists were placed in direct competition with new young scientists for both university jobs and federal support.

The plight of these older scientists constitutes a pressing short-term national problem; the effects and implications of the age distribution problem are of much longer term. Inflation, which is reducing the current $7 million in annual federal support of atomic and molecular physics by about $500,000 each year, also is a long-term problem. Some attempt to compensate for this loss and to facilitate the entrance of a few young scientists each year into this field is necessary. A very serious decline in activity will begin in about five to ten years unless provision is made for the training of new young atomic and molecular physicists. A pause in expansion today will result in a reduction in the amount of available talent in five years, which is the time required for graduate training. An intelligent student who sees little opportunity in a field looks elsewhere; consequently, he will not be available as an atomic and molecular physicist in five years. Some temporary artificial substitute for the customary turnover of personnel must be provided if this field is to survive. A small but reasonable figure for expansion would be ten new scientists per year. If continued for a sufficient length of time, this annual increment would eliminate the age distribution problem by allowing a population growth of 3 percent. The cost would amount to an annual increment of about $400,000.

RECOMMENDATION 4 The federal government should attempt to provide an additional annual increase in funds to (a) reduce the effects of inflation and (b) allow at least ten young scientists to enter atomic and molecular physics each year. This increase should amount to approximately $0.9 million per year, $500,000 to mitigate the effects of inflation and $400,000 for the support of new investigators.

The NSF recently instituted major new efforts to support science related to national goals. Although this Committee agrees emphatically that every possible attempt should be made to bring modern science to bear on the problems facing society, we believe that those concerned with support policies should recognize that these new programs are not designed to meet the type of support problem discussed in this report. Although atomic and molecular science undoubtedly will play an important part in many of the projects funded by these new programs, the stated objectives of the RANN (Research Relevant to National Needs) program appear to be directed toward obtaining concrete results rather than toward expanding basic research capabilities and thus allowing basic science to make a greater contribution to national needs. The RANN program seemingly is designed to capitalize on developed scientific capability in the achievement of fairly specific objectives and should not be confused with programs concerned with basic research.

A gap exists between the support of basic research and of such programs as NSF's IRRPOS (Interdisciplinary Research Relevant to the Problems of Our Society); steps should be taken to narrow this gap. The funding of basic research programs is based on their worth as fundamental science. The funding of applied research programs, including IRRPOS, depends on their immediate relevance to some definite goal. However, much work that scientists must perform lies in the region between these two extremes—for example, work that entails the careful examination of physical processes and phenomena to gather data and improve the level of detailed knowledge in subject areas with the promise of practical application. Currently, this type of endeavor is neglected, especially since the occurrence of budget reductions in the DOD agencies and NASA and the concomitant project terminations by industry. Traditionally, these sources supported the type of exploratory research that is essential in converting new scientific knowledge into technology. At the present time, the effort to collect accurate data on broad categories of atoms and molecules is being abandoned. Formerly, the National Bureau of Standards did a great deal to coordinate the collection of atomic data by scientists in government, universities, and industry. This type of program should be revived and expanded. To be effective, such a program must provide for greater intercommunication among its participants and should seek fuller knowledge of both atomic and molecular processes.

RECOMMENDATION 5 The National Science Foundation should establish an effective method of supporting the research that is intermediate on the basic to applied continuum. Specifically, we *recommend* that the support of research in this intermediate area be closely coordinated with the efforts of the National Bureau of Standards, the National Institutes of Health, and other agencies having a strong interest in this type of research.

Finally, we must consider the particular needs of interdisciplinary laboratories and other types of large groups tantamount to national centers that attempt to relate the findings of atomic and molecular science to the needs of other disciplines and to practical problems of national concern. The Joint Institute for Laboratory Astrophysics of the National Bureau of Standards and the University of Colorado has become a major center for the examination of atomic and molecular processes of interest to astronomers and upper-atmosphere scientists. New York University, the University of Washington, Yale University, and the University of Pittsburgh have long maintained programs of unusual experimental and theoretical capability in atomic physics, some of which are directed toward processes occurring in the atmosphere. The Massachusetts Institute of Technology sponsors an especially active group working in basic and applied aspects of quantum optics and electronics. The University of Maryland has established a strong atomic and molecular physics program to study atomic, molecular, and fluid processes related to aeronomical, meteorological, astrophysical, and plasma phenomena. Efforts of this type constitute an important national resource. To assemble the variety and depth of talent that these groups represent is difficult. The personnel must be chosen carefully so that their interests interact and stimulate one another's thinking; individual ability and versatility must be much above average if interdisciplinary synergism is to occur. These groups, and a number of others not specifically mentioned, are extraordinarily productive and increasingly valuable. Unfortunately, most are experiencing great difficulty in obtaining the type of support that makes them particularly valuable.

Interdisciplinary basic research has not been emphasized in the NSF program, although the trend toward such research is strong in the universities. Although it is possible for two or more different offices within the NSF to join in the support of a single proposal, they rarely do so. Possibly, the funds for such an undertaking have to be available at the appropriate time in each participating section, which is increasingly unlikely to occur under present conditions. However, by its nature, interdisciplinary research is peripheral to the interests of any well-defined field. It is difficult to get a strong consensus among the reviewers of an interdisciplinary proposal, which puts such requests at a disadvantage in the competition for funds.

Larger group efforts, such as those just mentioned, face additional hazards. To be maximally effective, support must be sufficient to allow research mobility so that members of the group can explore new ideas together and develop new techniques. A certain amount of spontaneous interaction and free flow of ideas is exceedingly important for success; without this element a group effort cannot be meaningful. Long-term visitors also are of great value in contributing new ideas. Often they return to their institutions with a new outlook and new interests.

RECOMMENDATION 6 The National Science Foundation should give immediate attention to the effecting of greatly improved means of support for interdisciplinary basic research, apart from its several current programs in interdisciplinary applied research, particularly for large groups and major centers at universities, government laboratories, and industrial laboratories that are attempting to make a broad attack on complex research problems. The National Science Foundation also should make a special effort to ensure the continuity of established programs and interdisciplinary research groups that show promise of continued high productivity and aid in making these groups a resource for the scientific community in general.

Appendix I.A

American Physical Society
Division of Electron and Atomic Physics
1968 Survey of the Status of Atomic and Molecular Physics

FOREWORD:

The Bylaws of the DEAP set forth the domain of its interests as "the advancement and diffusion of knowledge regarding the electron, atoms and simple molecules, the scope of this Division shall comprise the fundamental principles of electron and atomic physics, physical phenomena involving electrons, ions, atoms and simple molecules, including their interaction with the electromagnetic field, and the application of these principles of phenomena."

For the purpose of this survey and in more operational terms it should include the interactions of electrons, ions, atoms, and molecules in the gaseous state and at surfaces. It includes their interactions with static electric and magnetic fields, with radiation and with each other through collisions. Work in such fields as astrophysics, aeronomy, plasma physics and others is to be included if the specific project is aimed toward obtaining knowledge about the basic processes themselves.

On the other hand, if the intent of a project is more nearly directed toward specific technological application it probably should be excluded. Those aspects of quantum electronics that are generally thought of as solid state physics should be excluded, as should statistical mechanics and plasma physics considered as many-body phenomena.

A case that is sticky to decide should be included, together with a notation indicating the nature of the classification difficulty and your suggested interpretation.

Some of the questions obviously pertain to the university scene. However, this circumstance should not be interpreted as a lack of concern for the situation in other kinds of laboratories. It is obvious that the future of any field is dependent on the production of new workers in that field, and special attention must be given to this aspect of the overall state of affairs.

The outline that follows is intended as a guide rather than a strict injunction. Remember, though, that quantitative conclusions are critical and therefore, information must be given in a form that is fairly consistent among the returns. Certainly, any additional information that you deem important, or any suggestions you or your colleagues care to make, will be most welcome.

INFORMATION REQUESTED:

1. *Basic Support*

   a. List all existing grants and contracts supporting atomic and molecular physics giving the Federal agency or other source, the senior scientists associated with each grant or contract and their immediate affiliations (e.g., Physics Department, Chemistry Department, etc.), starting and termination dates, the annual rate of support (total amount, # years) and a very brief title. How complete do you estimate your coverage to be?

   b. How many graduate students are associated with the total atomic and molecular physics research effort indicated above? Indicate how many are supported by grants or contracts, how many by fellowships and traineeships, and how many are paid as TA's due to lack of contract funds but are actively engaged in research.

   c. How many post docs (fractional time if appropriate)?

   d. How much money is budgeted for permanent equipment per senior investigator from grant or contract support (average amounts)? If this is being supplemented by internal funds, or by substantial amounts of construction of equipment that might better be purchased, estimate the additional (average) amount per investigator needed if grants and contracts were to assume this additional burden. Are the total permanent equipment needs temporarily higher or lower than average?

e. Is the total research support per senior scientist increasing, decreasing or remaining static over the past few years? Please be quantitative about this. Mention in Section 3 any significance this may have.

f. How certain is the funding of these projects for FY 68–69? On what federal agencies, if any, does this support depend?

2. *Additional Support Needs*

a. List scientists having no continuing Federal support who have 1) either submitted a proposal seeking such support within the past 18 months, or 2) are new colleagues (within 18 months) in the process of preparing a proposal and actively seeking support. Where did they come from? How many of these have completed their Ph.D. within the past 3 years? List any scientists who have been in your organization more than 18 months and who have been unsuccessful in attempts to gain support.

b. Estimate the amount of support needed for these scientists to establish research programs. Please do not be extravagant.

c. If possible, identify those scientists listed in Section 1 who have obtained support by associating themselves with an existing program that is being supported, but who are not able to do the type of research in which they are interested. Estimate the additional support required for these men to work independently.

3. *General Description*

Give a general discussion of atomic and molecular physics problems at your institution based on the following questions: Have additional scientists been acquired to work in this area within the past 18 months? Are they senior or junior? Are new scientists having to affiliate themselves with existing research programs for lack of outside support? Are important new programs not being undertaken for lack of support? Do you have definite plans to increase the number of scientists in atomic and molecular physics research and by how many? Is there evidence for a significant change in graduate student demand for research supervision in atomic and molecular physics, or from any other source? Is new permanent equipment a problem? What specific problems exist now, or can be expected as a result of either the character or the magnitude of Federal support of research? Please mention any other factors relating to the Federal support of research that are salient in assessing the current status of atomic and molecular physics research and problems relating thereto. What fields of atomic and molecular physics do you think will be especially demanding of support during the next few years?

II TECHNICAL REVIEW OF ATOMIC AND MOLECULAR PHYSICS

5 Introduction

Atomic and molecular physics occupies a central position in relation to basic and applied research in a large number of disciplines, as shown in Figure 4. In this Part we describe these relationships and present a summary of the present state of the field. We attempt also to indicate the potential of atomic and molecular research in the 1970's.

A. THE PHYSICS OF THE GASEOUS STATE

Atomic and molecular physics may be defined as that branch of research that is concerned with free atoms, molecules, and electrons and their interaction with each other and with electromagnetic radiation. However, because of the need for information about atoms, molecules, and electrons to solve problems in fields ranging from astrophysics to industrial processing, atomic and molecular physics has evolved into the study of the physics of the gaseous state. The physics of the gaseous state is no longer the exclusive province of those trained in physics. Because the delay between research and its applications is often brief in this field, chemists and engineers representing a number of disciplines have been attracted to it and are conducting

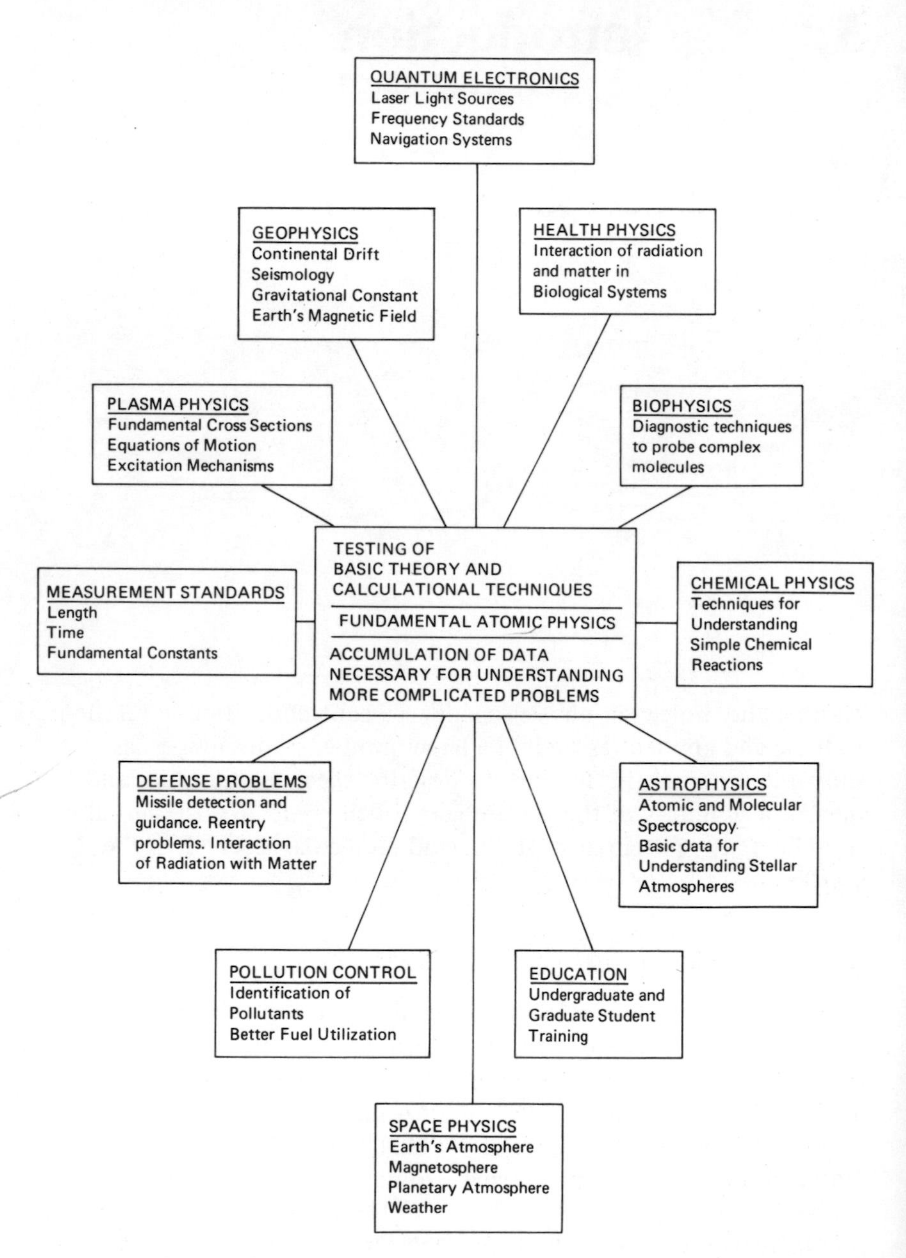

FIGURE 4 Relationship of atomic and molecular physics to other disciplines.

research in atomic and molecular physics. At the same time, the compelling need for atomic and molecular data to use in applications has become increasingly important in determining the types of research that are performed. The increasingly frequent inclusion of atomic and molecular physics courses in curricula of departments of applied science and engineering and the introduction of cross-disciplinary programs such as chemical physics further emphasize the close relationship of the basic science content of this field to applications in other disciplines.

Atomic and molecular physics significantly affects national decisions on some aspects of public policy. For example, the acceptability of a nuclear weapons test ban treaty depends on our ability to predict the effects of nuclear explosions without actual testing. This knowledge is obtained from theoretical models that use as input a wide variety of fundamental atomic and molecular data.

Other major policy decisions relate to the alteration of the earth's atmosphere and climate caused by human activities. Basic atomic and molecular physics has direct bearing on these problems. The urgent problem of smog production and control involves the interactions of solar radiation and molecules and the subsequent photochemical reactions. The longer-range problems of climatic change that may arise from an increase of the abundance of carbon dioxide and other more deleterious pollutants in the atmosphere and that alter earth's radiation balance are of unquestioned importance. The study of the atmospheres of the planets Mars and Venus, which consist mostly of carbon dioxide, suggests the possible future of the earth if wisdom is not exercised in dealing with our own environment.

Not all atomic and molecular physics is motivated by the need to solve practical problems. The field is rich in intellectual content and basic discovery; these features are primarily responsible for its present vitality. Nevertheless, its potential for application is a particular characteristic of contemporary atomic and molecular physics. This field has a long and proud history of giving new ideas, new technology, new industries, and new wealth to society.

At the turn of the century, experiments with electrons led to the discovery of x rays. Gaseous electronics experiments led to the development of many devices that are used in communications, electrical power production and distribution, and the illumination industry—for example, mercury and gaseous rectifiers, thyratrons, and fluorescent lights. Precise determination of ionic masses in the 1920's substantiated Einstein's theory and showed that nuclear energy could be released. Basic research on the energy levels of atoms and molecules was the *sine qua non* for the development of gas lasers. Ion implantation, photoelectricity, detectors for nuclear radiations, heat sensing—the list of powerful additions that atomic and molecular physics has produced is long.

Atomic and molecular physics is a "small science" in which research is done by individuals or small teams. Often the tools needed for the research must be invented in the course of the work; these tools usually are incorporated in related technology. Vacuum-production techniques, analytical instruments, gas-purification techniques, and sources of ion and neutral beams are examples. In chemical analysis, optical spectrometers, originally devised for research in atomic and molecular physics, are now found in countless industrial laboratories. Mass spectrometers are widely used in industrial laboratories and in production plants for process monitoring. Electron monochromators, invented within this decade for the study of resonances in electron scattering by atoms, are used in a new type of chemical analysis—Auger analysis—which holds unique promise for quality control in the manufacture of microelectronics elements for improved computers. The use of high-energy resolution electron scattering to obtain unique signatures for complex molecules provides a new and exciting diagnostic technique, which may offer serious competition to conventional mass spectroscopy. This type of rapid incorporation of devices from research laboratories into industries is common in atomic and molecular physics.

Research in atomic and molecular physics is usually conducted by small teams using low-cost equipment. The laboratories active in this field are widely dispersed geographically. They exist in universities, government establishments, nonprofit institutions, and industry. Atomic and molecular physics research is peculiarly well suited to small universities and colleges; it offers the double advantage that ideas and devices are contributed by people with a remarkable variety of independent viewpoints and that the basic research has direct relevance to the educational experience. The increasing cost of research, even in small science, threatens to change this situation.

B. A FRONTIER FOR RELEVANT RESEARCH

Physics possesses many intellectual frontiers. The challenge of atomic and molecular physics lies in achieving a detailed understanding of the properties of atoms and molecules, acting individually and in groups.

The interpretation of the optical spectra of simple atoms—that is, the structure of atoms—constituted one of the first problems in the modern era of physics. It was successfully attacked in the 1920's by the application of quantum theory. Most of the subsequent effort in atomic and molecular structure centered on high-precision measurements of atomic and molecular spectra. With the advent of infrared and radio-frequency techniques, which provided information about chemical bonding and molecular vibrations accompanying energy absorption, molecular physics became an important part of chemistry. Molecular vibration is critical to the understanding of the transfer of energy through gases. The inventions of the maser and laser came directly from a study of the vibrational states of ammonia molecules. The invention of the atomic clock was a direct outgrowth of the study of radio-frequency spectroscopy of atoms.

Atomic and molecular physics contributed heavily to World War II technology. Advances during that period stimulated further research after the war ended. Radar, fast electronics, and vacuum techniques are among the fruits of the wartime efforts in atomic and molecular physics.

The massive electronics and communications effort of the Department of Defense was not terminated entirely after World War II. Due to the vision of many of the scientists and administrators involved, particularly in the Office of Naval Research, some programs remained that attempted to maintain a basic research effort, primarily directed toward the practical goal of generation, propagation, and detection of electromagnetic energy.

In the postwar period, new agencies such as the Atomic Energy Commission (AEC) and the National Aeronautics and Space Administration (NASA) were established, which, with the Department of Defense, financed basic research efforts, and several outstanding university laboratories flourished. These activities resulted in new methods for dealing with the complicated interactions of atoms, electrons, molecules, and ions in gases as well as the initiation of research on a more complex but more readily applied plane.

One approach to the discovery of the dynamic processes underlying chemical reactions is through the study of colliding beams of particles—atoms, ions, molecules, electrons. The idea of such crossed-beam experiments is not new, but many years were required to develop the techniques and to perfect the scattering-beam experiments. It was necessary also to develop precise techniques for data analysis. Little could be accomplished before the advent of large and fast-pumping vacuum systems; it is no coincidence that the field of vacuum technology experienced its most rapid growth during the past decade. Usually the energies of the interacting particles are low, and the problem of generating beams of such particles is a difficult one. The detection of scattered uncharged atoms and molecules is still more difficult. In many cases, particularly in data-processing techniques, lessons gleaned from recent developments in high-energy nuclear physics have proved invaluable.

Because of the exciting new opportunities available for exploiting their arsenal of sophisticated experimental techniques, nuclear physicists are entering atomic and molecular physics. Atomic physicists also are using the tools of the nuclear physicist. For example, Van de Graaff accelerators are employed in many applications involving particle scattering, beam-foil spectroscopy, channeling, and coincidence techniques.

Research in atomic and molecular physics is clarifying and elucidating problems in chemistry and the atmospheric sciences. Phenomena generally possess no single physical explanation. Chemical and gaseous phenomena are made up of a variety of simultaneous physical processes, and each of these must be examined in detail so that its role can be properly assessed. Solutions to such problems ordinarily require years of careful research. An exciting aspect of the field is that tools are evolving that allow sophisticated and penetrating investigations.

The frontier of socially relevant physics includes problems in engineering, chemistry, biology, and atmospheric sciences that, a few years ago, would have appeared to be too complex to attack from a fundamental viewpoint, but the solutions of which require fundamental knowledge derived from basic research. Plasma physics, which is the study of the behavior of charged particles, ions, and electrons in the gaseous state, moving under the influence of applied and self-generated electromagnetic fields, provides one of the most striking examples. In recent years plasma physics has emerged as a vital part of physics, closely related to atomic and molecular physics. Plasma scientists are pursuing new methods of chemical synthesis and power production and are active in the study of upper-atmospheric effects and astronomical effects as basic research in physics. Accurate simulation of the ionosphere and the magnetosphere of the earth may soon be feasible in the laboratory. The interaction of the solar wind, a stream of protons from the sun, with the ionospheres of Mars and Venus may cause significant modifications to the atmospheres of the planets. In its interaction with the magnetosphere of the earth, the solar wind is a major factor affecting the reception of solar energy by the earth and may be an important element in weather processes.

Plasma physics is concerned also with the macroscopic behavior of fluids—fluid mechanics. This complex field deals with the behavior of a fluid, liquid, or gas as it moves or as foreign bodies move through it. Detailed descriptions of the fundamental atomic interactions are the basic input to fluid theories although the many-body nature of the fluid adds an additional essential new aspect to the problem.

Plasma physics is perhaps the best known recent example of the relation of atomic and molecular physics to frontiers of relevance, but there are many other examples. Lasers are used with increasing frequency as monochromatic light sources for the study of photochemistry. Recent measurements bring important new understanding of processes involving intermolecular chemical reactions. The physics of fluids is now examined on an atomic and molecular level and is beginning to shed new light on life processes in fluid media. In addition, it provides a deeper understanding of the microscopic aspects of fluids.

The development of atomic and molecular beam techniques was among the most significant advances of recent years. Molecular physicists and physical chemists use these techniques to examine a wide variety of basic chemical reactions and energy-exchange processes, including those relating to systems of biological and genetic interest.

The history of the development of atomic and molecular coherent radiation sources—of masers and lasers—is well known. Infrared and visible laser light sources are both products and new frontiers of atomic and molecular physics research. This challenging new field originated in molecular beam spectroscopy and measurements of atomic collision cross sections and optical lifetimes and transition probabilities; its future development and growth depend on the continued progress of these kinds of research.

As a final illustration, we mention the recent spectacular growth of molecular astronomy. This new field, in which new molecules are being discovered in interstellar space almost weekly, possesses the deepest implications for our understanding of the origins of life processes and cosmology. Yet the objects of molecular astronomy are only the common molecules, large and small, that have been studied in laboratories for years.

The applications of basic science grow continuously as we achieve a deeper understanding of the behavior of large ensembles of particles in gases, liquids, plasmas, and complex molecules. Problems that previously have not been amenable to solution may soon be subjected to theoretical analysis and experimental investigation that will lay the groundwork for a new role of basic science in meeting human needs.

6 Outline History of Atomic and Molecular Physics

In assessing the current role of atomic and molecular physics, some aspects of its history must be considered. It should be noted, however, that this is not intended to be a comprehensive historical review. The emphasis is on the development of atomic and molecular physics in the United States. The subject possesses a notable history in other countries as well—this is only briefly touched on here. From the beginning of the century to the mid-1930's, atomic physics was the center of growth for modern physics. In the late 1930's the most active and innovative atomic physicists turned very largely to the new problems of nuclear physics, and their conversion to nuclear physicists became complete during the war years. After 1945 atomic physics (especially collision physics) was completely out of fashion as an academic discipline. Its rebirth, beginning slowly about the 1950's, came about not solely because of academic interest but rather because of pressing needs in a variety of applied problems: industrial, military, atomic energy, and space. Responding to these applied problems, the field has bloomed again, producing many innovations of great scientific significance as well as important technological advances. Belatedly, academia has recognized the renewed scientific importance of this field by welcoming it again in faculties of physics, applied physics, and related departments. A second reason for the slow rebirth of atomic and molecular physics was that the very difficult experimental techniques that are often required were not fully developed until quite recently.

In the following partial survey of the history of recent developments in atomic and molecular physics, we may find important lessons for the future in that the rebirth of this science was stimulated principally by practical needs and that the satisfaction of these practical needs required and elicited major developments in basic science.

The first phase in the history of atomic and molecular physics was initially dominated by spectroscopy, which contributed so much of the experimental data leading to the development of quantum mechanics. Work in electron and ion beams, particularly involving mass spectroscopy, was also important during this growth phase, as was the study of the Zeeman effect and of spatial quantization, so dramatically demonstrated by the Stern-Gerlach experiment. In addition, the study of atomic and molecular collision processes, to which the Göttingen group led by J. Franck contributed so much, developed rapidly during the 1920's. The low-energy electron collision experiments by C. Ramsauer, also in Germany, and the electron swarm experiments pioneered by J. Townsend at Cambridge during the 1920's greatly stimulated interest in applying the newly developed ideas of quantum mechanics to scattering problems. Quantum mechanics also was rapidly applied to explain the foundations of chemistry, both in molecular structure and, with the work of F. London, M. Polanyi, O. K. Rice, and H. Eyring, in the theory of chemical reaction rates.

By the mid-1930's, a successful foundation was established in all these fields, and a strong and continuing growth of atomic and molecular physics was anticipated. This progress was interrupted largely by historical accidents. First, the rapidly developing knowledge of nuclear reactions led many of the atomic physicists to apply recently developed techniques to this new class of problems. In both experiment and theory, nuclear collisions soon became dominant, and the work in atomic beams and spectroscopy was exploited in the study of nuclear moments. Second, many atomic physicists began working on nuclear problems during World War II, and others went into electronics and radar; as a result, atomic physics research lost momentum.

In addition to these historical accidents, many people in physics felt that the major discoveries of atomic physics had already been made and that the field had little more to contribute. A prevalent viewpoint was that all chemistry and all the physics of atoms outside the nucleus could now be understood in terms of the Schrödinger equation, the symmetries of the particles (electrons and nuclei), and the known electromagnetic forces; only tedious problems of computation remained, as all the challenging problems had been solved.

This point of view did not diminish the appreciation of activities such as the development of solid-state physics, but it did very much affect the way in which atomic physics was treated in the years after 1945. Undeniably, at this time the major areas for exploration of the unknown were in nuclear physics and high-energy physics. However, as subsequent events have shown, the collective interactions of several particles, as they occur in atoms, molecules, solids, superfluids, or plasmas, continued to produce qualitatively new phenomena, providing nontrivial problems for experimental study and challenging the theoretician to develop the concepts needed for their intuitive understanding and the computational techniques required to explain them quantitatively. But the notion that atomic physics was obsolete retained a tenacious hold on the minds of most physicists. As late as 1958, when the celebrated text on quantum mechanics by Landau and Lifshitz[4] appeared in English translation, a well-known physicist gave it a very complimentary review and then suggested that in the preparation of future editions "magnetic field effects would well be expanded at the expense of the two chapters on molecular physics. ... Probably more people are interested in the former topic than in the latter."[5]

During the period 1938–1945, wartime activities affected the direction of subsequent work in atomic physics in many ways other than deflecting the activities of most physicists to other problems. The two largest wartime physics undertakings, the electronics and radar work centered at the Massachusetts Institute of Technology (MIT) and Harvard University and the nuclear bomb project, drew extensively on existing knowledge of atomic and molecular physics. Simultaneously, they created a vast array of new equipment and techniques that were available after the war for use in many fields, including atomic physics.

Two main parts of atomic and molecular physics are resonance physics and collision physics. In resonance physics we include mag-

[4] L. D. Landau and E. M. Lifshitz, *Quantum Mechanics* (Addison-Wesley Publishing Company, Reading, Mass., 1958).

[5] M. E. Rose, *Physics Today*, *11* (No. 12), 56 (Dec. 1958).

netic studies, atomic beams used for the study of nuclear moments or molecular magnetic moments, radio-frequency and microwave spectroscopy, masers, lasers, optical pumping, and, with a slight extension of the usual usage, classical optical spectroscopy. Collision physics includes the collisions of atoms, electrons, ions, and small molecules, both in single collision processes and in bulk gases or plasmas. It also includes the collisions of isolated particles with the central cores and inner shells of atoms in condensed media as well as collisions with atoms on solid surfaces.

Resonance and collision physics had rather different histories in the years following World War II. Resonance physics developed primarily in university circles where it remained a comparatively small but accepted part of standard physics departments. This work was sometimes brilliant, a notable example being the atomic beams research conducted by a group at Columbia University under the leadership of I. I. Rabi. In general, atomic-beam experiments were devoted principally to the study of nuclear moments and to various magnetic interactions in atoms and molecules. This field profited directly by the advances in electronic and radio-frequency techniques that resulted from the radar project at MIT and Harvard during World War II. Radio-frequency spectroscopy was applied to atomic and molecular beams, making measurements of increasingly high precision possible. One of the most important results of this work was the discovery of the Lamb shift, with its implications for quantum electrodynamics.

The development of radar and the techniques that arose from it led to the discovery of nuclear magnetic resonance, made independently by E. M. Purcell at Harvard and F. Block at Stanford. Not only did research on nuclear magnetic resonance develop rapidly, but it was soon used in chemistry as a powerful technique for the analysis of structure and interatomic interactions. Closely related to nuclear magnetic resonance were studies of electron spin resonance, which also were of particular importance in chemistry.

The postwar work in microwave spectroscopy that developed at several institutions, including Harvard and Duke, also employed the techniques and surplus equipment available for radio-frequency studies. This field has led to the development of a strong branch of chemistry associated particularly with rotational interactions in molecules.

Somewhat independent of the preceding research efforts in resonance physics is optical spectroscopy. Work in this older field continued in university surroundings as a modest component of the research picture in the post-World War II period. Among the workers who continued the early traditions of optical spectroscopy were G. H. Dieke, R. S. Mulliken, G. Herzberg, J. E. Mach, and A. H. Nielsen. However, the major postwar interest in spectroscopy was more in chemistry than in physics. Atomic spectroscopy was generally ignored except for those circumstances in which it was needed as a diagnostic tool, and it largely disappeared from academic curricula. Because of applied demands, the National Bureau of Standards developed one of the great centers in spectroscopy. With such varied developments as atomic clocks, lasers, and beam-foil spectroscopy, renewed life came to many portions of spectroscopy.

The optical laser is, of course, one of the most significant new developments of recent years. Both a product and a prime source of atomic and molecular physics research, its history is too complex to present in detail. A major precursor of laser development was optical spectroscopy, which contributed not only the optical techniques necessary to obtain laser action but also the basic input of energy levels, line shapes and strengths, and optical radiation theory so necessary to the development of lasers. The somewhat earlier development of the maser by C. H. Townes and his students at Columbia University was the other essential element in the sequence of events leading to the construction of the first laser, by T. Maiman at Hughes Research, and the first gas laser, by A. Javan and W. Bennett at Bell Telephone Laboratories. The development of ammonia and atomic hydrogen masers was stimulated to a large extent by the magnetic and electric resonance absorption and beam spectroscopy, which had their origins in the magnetic resonance work of I. I. Rabi and his students at Columbia.

In the brief period of ten years since the construction of the first laser, the field has grown to gargantuan size and is continually demonstrating its ability to assist in the solution of both fundamental and practical problems. Chapter 10 discusses some of the more direct connections between laser physics, laser technology, and atomic and molecular physics.

Because resonance physics and spectroscopy stimulated widespread academic interest, work in these fields was conducted to a large extent in universities. Particularly in the first ten years after World War II, the level of effort in these fields of physics was relatively small in comparison with the substantial efforts in nuclear and high-energy physics and in solid-state physics. However, it proved fortunate that small-scale efforts in resonance physics and spectroscopy continued in universities.

The situation in the other major field of atomic and molecular physics, collision physics, contrasted strongly with that in resonance physics. In the prewar years, atomic collision physics had been closely connected with spectroscopy, particularly in the work of such groups as that at Göttingen led by J. Franck. Atomic and molecular collisions were obviously important to astronomers and spectroscopists because of their role in exciting and de-exciting spectral lines. In the postwar period, Franck continued some of this work at The University of Chicago, but a small group under O. Oldenberg at Harvard became a more active center for this type of research. Oldenberg, in particular, trained a number of students who played a major role in founding the next generation of collision physicists. Many of these students, however, made their important contributions not in universities but in industry or government laboratories.

Much more important in the postwar years as a field of research in collision physics was gaseous electronics, including discharge phenomena, arcs, swarm studies, and the like. This field had played a part in the electronics and radar project during World War II. MIT, under the leadership of W. P. Allis and S. C. Brown, remained an important center for gaseous electronics, which was related in a very significant way to work in the broad field of electrical engineering. Another major center for research in gaseous electronics was the University of California, under the guidance of L. G. Loeb. In spite of the major contributions of these two academic centers, the main focus of activity in gaseous electronics in the postwar years was in industry, including such major laboratories as that of General Electric at Schenectady, Bell Telephone Laboratories, RCA, and Westinghouse. Perhaps the most important center for the development of this field in the next 20 years was that established in the early postwar years at Westinghouse under the direction of D. Alpert. Alpert collected a major team of scientists in this field, including T. Holstein, R. Fox, M. A. Biondi, G. J. Schulz, A. V. Phelps, and others, to study the fundamental processes underlying the important practical problems of gaseous electronics, arcs, and discharges that were of importance to the electrical industry. This group was quick to recognize the major technological obstacles to continuing advance in the understanding and solution of the many scientific problems of this field. Among these was the practical one of the limitation of vacuum technology to about 10^{-6} Torr. A major effort was devoted to the development of techniques for producing, measuring, and maintaining considerably higher vacuum. Among the products of these efforts were the Alpert valve and the Bayard-Alpert gauge, which became for many years essential tools in vacuum technology in the 10^{-8} and 10^{-9} Torr pressure range. Because unwanted collisions with ordinary background gases had been such a severe limiting feature to atomic collision studies, this development made possible vastly accelerated advances in many areas of atomic physics and many laboratories.

A second engineering-oriented application of molecular collision phenomena, which has played an important role in atomic and molecular physics since World War II, is the general field of fluid dynamics, shock waves, and jet propulsion. Collision phenomena in hot gases were of concern in connection with military ordnance (in both interior and exterior ballistics) during and after World War II. They were also important in aerodynamics, jet propulsion, rocket phenomena, and, ultimately, phenomena associated with spaceflight. The study of shock waves in particular was greatly accelerated during World War II in connection with both conventional weapons and nuclear phenomena and nuclear weapons. In the postwar period these lines of work were conducted mainly in departments of aeronautical and mechanical engineering, the aerospace industry, or government laboratories such as that at Los Alamos or those of NASA. Important engineering laboratories loosely associated with universities, such as the Jet Propulsion Laboratory of the California Institute of Technology, Cornell Aeronautical Laboratory, and the Stanford Research Institute, performed major research in shock waves and energy transfer phenomena in gases. The connection of these phenomena with chemical kinetics led to the development of this kind of experimentation in some chemistry departments, notably under G. Kistiakowsky at Harvard and S. Bauer at Cornell. In the post-Sputnik era, increased concern for these and other areas of collision phenomenology resulted in a great and rapid increase of effort in the aerospace industry and in laboratories in many areas under the sponsorship of the Department of Defense (DOD) and NASA.

Also important for the application of atomic physics in the immediate postwar years was its role in nuclear experimentation. Processes and techniques involving phenomena of atomic physics had implications for work in such diverse fields as ion sources for accelerators; detection devices for charged particles; the understanding of stopping powers, energy loss of charged particles, unwanted collisions with background gases, and excitation processes leading to energy loss by radiation or ionization; and the study of the ultimate damage produced by energetic particles passing through matter of various kinds, including biological materials. For these reasons, work on radiation physics (as well as radiation chemistry) was conducted at several of the national laboratories under the auspices of the AEC, and similar work also was supported in other institutions. The studies of the processes resulting from the passage of energetic charged particles through matter by S. K. Allison at Chicago and T. Lauritsen at Cal Tech are notable exceptions in the dearth of academic work in nonnuclear collisions in the postwar years, although, of course, these studies were made because of nuclear needs.

With the development of Project Sherwood and the attempt to develop nuclear power by fusion instead of fission, there arose a new need for considerable information and new techniques associated with atomic and molecular collisions. This subject is discussed in some detail in Section A of Chapter 11.

From a very early period, the propagation of radio waves in the atmosphere, and particularly the phenomena of the ionosphere, led to a recognition of the importance of free electrons in certain regions of the atmosphere. Important electron collision problems were associated with the loss of electrons and also with other interactions. In the postwar period problems of radar propagation and of missile guidance, detection, and ranging added impetus to the study of some of these collision processes associated with the propagation of electromagnetic radiation. Much of this work had obvious military applications and was sponsored by the DOD. These problems, and those of nuclear-weapons effects on the atmosphere, led to a great deal of fundamental work on collisions that was supported by DOD.

The basic work in chemical reaction rates took place in the early 1930's, including that of F. London, M. Polanyi, E. P. Wigner, H. Eyring, and others. When basic studies of chemical kinetics were resumed in the postwar period, the leadership of atomic physicists, such as E. P. Wigner and H. Bethe, was missing, and it was a long time before it was replaced by a new generation of collision theorists and collision experimenters derived from the chemical community. Meanwhile, the major emphasis in chemical kinetics was not in pressing forward in basic collision theory but rather in exploiting the advances that were made in the 1930's and applying them to more complicated reacting systems. Although the application of collision theory to the structure of liquids and gases was effected under J. O. Hirschfelder at the University of Wisconsin, the great advances in the understanding of chemical kinetics in terms of collision processes really began with the experimental work on chemical reactions as a function of angle scattering, which was initiated by the experiment of S. Datz and E. H. Taylor[6] at Oak Ridge in the middle 1950's. This decisive work was done at a government laboratory, and in this case chemical physicists in a number of universities quickly took up the work and initiated the subsequent great expansion of the field.

Similar events occurred in mass spectrometry. In the 1920's and 1930's, mass spectrometry became a major research area of nuclear physics. In the 1940's and 1950's, it achieved central status as a chemical analytical tool, with applications to the analysis of complicated mixtures that were most fully realized in the oil industry. Some of the most important work on ion–molecule reactions was initiated at Shell Development Laboratories by D. P. Stevenson in the early 1950's. In principle, ion–molecule reactions are similar to chemical reactions when studied as single collision processes, except that one of the partners is charged and, in many cases, much easier to detect. In practice, the experiments usually were conducted independently by other workers and in other laboratories. Ion–molecule reactions are now studied in chemistry and physics departments and at government and industrial laboratories throughout the country.

The initial slow growth of the field of atomic collisions in the academic community (with the exception of such groups as those at Harvard, MIT, New York University, University of California at Berkeley, University of Chicago, and University of Pittsburgh) was striking. In the last few years, this situation has changed. A number of people have left industrial and government laboratories to accept university positions, and collision physics is becoming more common in physics departments than it was previously.

The situation in the United States in collision physics is very different from that in Great Britain. No review of this field in the United States in the past 20 years can ignore the tremendous influence of the strong Department of Physics at University College, London, under the guidance of Sir Harry Massey, which was and still is devoted very largely to aspects of collision physics and its applications in

[6] E. H. Taylor and S. Datz, "Study of Chemical Reaction Mechanisms with Molecular Beams. The Reaction of K with HBr," *J. Chem. Phys. 23*, 1711 (1955).

space research. This department excelled in the theory of atomic and molecular collisions, but it also had a strong experimental group associated with it. It was for many years the chief location for the development of atomic collision theory, not only in the English-speaking countries but in all the Western nations. As a consequence of its strength and productivity, collision physics is a major field in a number of British universities. In Northern Ireland, under D. R. Bates at Belfast, a similar group has developed that has extremely strong ties to the group at University College and also to many groups in the United States. Had it not been for the work of these groups in the British Isles, atomic collision physics would have progressed far less than it has.

Finally, mention must be made of the current vitality of the field in other countries, with noteworthy groups active in The Netherlands, France, Italy, Australia, Japan, and many Eastern European countries. For example, we note the effort in optical spectroscopy, optical pumping, and related subjects in France under the distinguished leadership of A. Kastler and J. Brossel. Perhaps the greatest number of active centers in atomic and molecular physics in European countries exists in Germany and the Soviet Union, both of which support many groups including some of great distinction.

7 Atomic and Molecular Physics as a Basic Science

One can think of atomic and molecular physics as consisting of two parts; the first of these, fundamental atomic and molecular physics, deals with the use of atoms and molecules as means of testing fundamental concepts in quantum mechanics and quantum electrodynamics. The second is the study of the many-body quantum system consisting of electrons and atomic nuclei—the atoms and molecules of our environment and the ways in which these systems interact with each other.

In this chapter, we attempt to illustrate the part that atomic and molecular physics plays in these vital areas by outlining some recent discoveries that have stimulated new understanding and some laboratory studies that provided crucial tests of theoretical advances.

A. FUNDAMENTAL ATOMIC AND MOLECULAR PHYSICS

The major theoretical innovation of twentieth century physics is the formulation of quantum mechanics and the subsequent quantization of the electromagnetic and electron–positron fields. The theory was invented to explain a variety of previously incomprehensible effects observed in atomic physics. Its success in explaining such phenomena as the atomic hydrogen spectrum, the diffraction of electrons by crystalline solids, and the spatial quantization of atomic magnetism, among many others, is one of man's greatest intellectual achievements.

Low-energy physics is now concerned with precise measurements on free electrons, free muons (heavy electrons), hydrogen-like atoms, and a few more-complex atoms and molecules. The motivation is the more precise testing of the basic assumptions of quantum electrodynamics and improving our understanding of the basic symmetries found in nature.

Quantum electrodynamics raises a number of conceptual problems and does not constitute a final or complete theory. It cannot explain or predict the values of the fundamental physical constants of nature, such as the mass of the electron and the muon or the speed of light. To achieve a complete theory, we must probe deeper into nature to uncover the smallest discrepancies in quantities that can be calculated from existing knowledge of the electromagnetic and electron–positron fields. From such work (the "low" road of precise measurement and small energy differences), together with efforts at the other extreme of high-energy and high-momentum transfer (the "high" road of billion-electron-volt accelerators, storage rings, and the like), the boundaries of knowledge of the basic nature of our atomic and subatomic universe can be extended.

The most important specific tests of quantum electrodynamics involve the comparison of theory and measurement of the gyromagnetic ratio (the ratio of magnetism to angular momentum) of the free electron and muon, the Lamb shift (small deviations from the Dirac theory in observed energy levels in hydrogen and hydrogen-like ions), the hyperfine structure (effects due to the intrinsic properties and structure of the interacting particles) of positronium, and the hyperfine structure of muonium (an atom consisting of an electron and a muon). The measurements of the hyperfine splitting of muonium are in qualitative agreement with theoretical predictions. Since agreement between theory and experiment for the Lamb shift and the magnetic moment of the electron is the basic evidence for the validity of quantum electrodynamics, there are some currently unresolved difficulties. With present technology, and with expected advances in the next few years, all these measurements can be significantly improved, which is one of the major objectives of atomic and molecular physics.

Atomic and molecular physics provides a means of testing the basic symmetry principles and invariances of physical laws. An example is the measurement of an upper limit to the electric dipole moment of the cesium atom. Under strict application of time-reversal symmetry and parity conservation in electromagnetic interactions, the electric dipole moment (electric charge times separation between positive and negative charge centers) should be identically zero. The upper limit is now established as equal to the electronic charge times a distance of 5×10^{-22} cm, thereby demonstrating the validity of time-reversal symmetry and parity conservation with surprising exactness. Other precision measurements have explored restrictions on the concepts of charge equality, charge conjugation (equivalence of electron and positron), and charge conservation, as well as provided further tests of time reversal and parity conservation in electromagnetic interactions.

In addition to values for the Lamb shift, measurements of the fine-structure separations in hydrogen provide values for the fine-structure constant correct to a few parts per million. A dramatic illustration of the basic unity of physics results from comparison of these with values obtained from a recent level-crossing experiment (resonance fluorescence) and from a combination of a Josephson effect measurement in superconducting solids and the values of the speed of light, the Rydberg, and the gyromagnetic ratio and magnetic moment of the proton.

Recent results on the Lamb shift confirm the existence of contributions from vacuum polarization, caused by the formation of virtual electron-positron pairs, virtual self-electromagnetic interactions of the electron, finite nuclear size effects, and relativistic mass corrections. The small discrepancy presently existing between theory and experiment in H, D, and He^+ suggests that a new assessment of theoretical predictions is needed. One possible source of the discrepancy is an underestimation of the proton-size contribution to the Lamb shift, which provides an example of the way that an atomic measurement can lead to a re-examination of the structure of the proton, one of the few stable particles in nature.

The most precise measurement extant in physics is that of the hyperfine separation in atomic hydrogen (the energy difference due to the two possible values of the magnetic interaction between the proton and the valence electron). The hyperfine separation, measured using a unique and ingenious maser cavity, is 1,420,405,751,786,4(17) Hz! The precision of the theory is sufficiently high that it must take into account the effect of proton polarization, that is, of the distortion of the charge distribution within the proton due to the proximity of the perturbing electron. This quantity, which is fundamental to elementary-particle physics, is related to the high-energy forward-spin-flip Compton scattering of the proton and to the neutron–proton mass difference. Its calculation requires consideration of the composite and resonant structure of the proton. Thus the atomic-physics measurement (together with the determination of the muonic hyperfine structure splitting) yields essential information on hadron (heavy particle) physics.

Positronium is an atomic system that should be determined entirely by quantum electrodynamics. An extensive analysis involving a suitable formalism (the Bethe-Salpeter equation) is required to match the experimental precision. The energy levels of positronium provide a severe test of electrodynamics and can yield a precise determination of the fine-structure constant. The measurement of the Lamb shift for positronium will be a definitive test of the theory, because the shift is entirely due to field-theoretic effects. Of fundamental value also is the measurement of the three-photon decay rate of *ortho*-positronium, now known to an accuracy of about 2 parts per thousand. Light-by-light scattering (the nonlinearity of the electromagnetic field) must be included in the calculation of the rate to this order. Thus atomic measurements provide a quantity to which photon–photon scattering makes a significant contribution.

The muonic atom, consisting of a muon and electron, is similar in many respects to the hydrogen atom, despite its short lifetime, but it is free of problems of hadron structure. The hyperfine structure of muonium offers a way of testing quantum electrodynamics and the Bethe-Salpeter equation, but more important, muonium could be the simplest system involving the muon and electron. The study of its energy levels might indicate whether there are any basic differences in the interaction properties of electrons and muons—a fundamental question of elementary-particle physics. Present findings suggest that the muon is only a heavy electron.

An illustration of the way that molecular physics affects experiments in quantum electrodynamics is the comparison of the measured hyperfine structure of the muonium ground state with theory. A precise value of the free proton to free muon magnetic moment ratio is required. The measured ratio must be corrected, since the chemical environment of a muon in water (or aqueous HCl), in which the muons are stopped, is different from that of the proton; therefore, the diamagnetic shielding correction also differs. In water this chemical shift reduces the applied magnetic field on a proton by 26 parts per million. However, because of its lighter mass and the resulting higher zero point energy, the muon forms a different type of bond. Rather than displacing a proton and entering into an H_2O bond, it is more likely to sit in the intermolecular space with less shielding of the applied field, giving a correction of roughly 10 parts per million. Here again we have an illustration of the relationship of atomic and molecular physics to particle physics.

Muonic atoms and molecules, in general, provide important fundamental information to other areas of physics. The muonic molecules consisting of two protons and a muon; a proton, deuteron, and muon; and two deuterons and a muon (systems formed when muons are slowed in water and heavy water) are interesting to study as nontrivial three-body systems in which the forces are known. In addition, a perturbation to the energy levels arises when the nucleons overlap due to strong interactions; fusion is possible. The muon capture rate in matter is needed to determine the weak-interaction coupling constant.

Many energy levels of heavy muonic atoms with heavy nuclei such as μ-bismuth or μ-lead have been measured with sufficient precision to determine nuclear properties, including accurate charge distributions, checks of nuclear polarization effects, and isotope, isotone, and isomeric shifts. In addition, the vacuum polarization due to electron–positron pairs is checked in the high-momentum transfer region, and Lamb shift effects, due to self-energy corrections to the muon, are on the threshold of detection. For many years the precise measurement of the $3D_{5/2}$–$2P_{3/2}$ level in muonic phosphorus (by the method of x-ray critical absorption) yielded the most precise value for the mass of the muon.

B. SIMPLE ATOMIC SYSTEMS

The specific quantum-mechanical problem that bridges the gap between fundamental atomic and molecular physics and the many-body problem is that of the three-body system. Such a system, in which the laws of force are known, would be completely susceptible to theoretical analysis. By providing difficult but soluble problems in perturbation theory these exercises are useful not only for stationary bound-state situations but also for dynamic scattering situations. Three-body problems provide a testing ground for computational procedures.

The problems include the calculation of eigenvalues for excited states, relativistic shifts and splittings of levels, Lamb shift, hyperfine structure, finite nuclear-mass effects, diamagnetic susceptibility, nuclear magnetic shielding, electric polarizability, nuclear quadrupole shielding, transition probabilities or oscillator strengths, scattering phase shifts, and photoionization cross sections. These calculations are necessary preliminaries to the understanding of complex atoms and scattering problems.

The simplest bound-state systems are those of the helium atom and the negative hydrogen ion. The basic scattering problems involve electrons, positrons, atomic hydrogen, singly ionized helium, and photons.

The processes that occur in simple systems are exceedingly diverse. Some examples are

radiative recombination, $H^+ + e \rightarrow H + h\nu$;
two-photon electric dipole decay, $H(2s) \rightarrow H(1s) + h\nu_1 + h\nu_2$;
single-photon magnetic dipole decay, $He(2^3S) \rightarrow He(1^1S) + h\nu$;
charge transfer, $H^+ + H \rightarrow H' + H^+$;
photodetachment, $H^- + h\nu \rightarrow H' + e$; and
electron-impact ionization, $e + H \rightarrow e + H^+ + e$.

Of these processes only the latter two have been the subject of precise experimental investigation; but even for them, greater resolution over a more extended energy range is required for theoretical comparisons.

Gross uncertainties occur in energy-level data. For example, there are two values, differing by 15 Å, for the wavelength of the 2^1S–2^1P transition of C V. The issue has now been resolved by theoretical calculations of the relativistic and radiative corrections.

The discovery of the threshold laws for electron impact excitation of atomic hydrogen demonstrates the value of studies of simple systems. Findings show that the cross sections are not zero at threshold. To the extent that other heavier atoms are hydrogen-like, their threshold laws also must be modified. These laws have implications for plasma physics and astrophysics.

C. COMPLEX ATOMS AND MOLECULES

To generalize the techniques and the results derived from simple atoms to the many-body atomic and molecular systems is difficult. There are no simple scaling laws from which two- and three-body results can be extended to many-body systems.

It is convenient to divide the study of complex atomic and molecular systems into two areas: the physics of atomic structure, or atomic spectroscopy, and the physics of atomic interactions or atomic collision processes. The first category includes properties of isolated atoms and molecules. Theoretical problems relate to the construction of solutions to the Schrödinger equation that take into account the Coulomb attraction between the electrons and the atomic nucleus and the mutual repulsions between the electrons. The experimental problems of the study of isolated atoms are most commonly the measurements of the electromagnetic radiations that are emitted and absorbed but also include studies of Auger electrons and of electron emission from autoionizing states.

Spectroscopy encompasses all regions of the electromagnetic spectrum. In practice, atomic spectroscopy covers frequencies varying from zero (the study of the interaction of atoms with constant electric and magnetic fields) to values surpassing 10^{20} Hz, corresponding to electronic transitions of the core electrons of heavy atoms. This range of over 20 orders of magnitude in frequencies demands expertise in many separate technologies, from the measurements of direct-current fields, through the various radio-frequency domains, through microwave, millimeter wave, infrared, visible, ultraviolet, and x-ray wavelengths. Each domain probes a characteristic aspect of the atom; all domains need to be investigated to achieve a comprehensive understanding of atomic structure.

Much of our knowledge of gross atomic structure, the results of nearly a half-century of careful experimental and analytic work of visible and infrared spectra, is summarized in the *Atomic Energy Levels* and Multiplet Tables, published by the National Bureau of Standards.[7] However, our knowledge of atomic structure is far from complete. Some of the more critical deficiencies are described in Chapter 7, Section D.

A parallel effort, in which known atomic transition probabilities for light elements of the periodic system are summarized, has also been published by the National Bureau of Standards.[8] Transition probabilities describe the lifetimes of excited energy levels; such lifetimes govern many important processes, including laser operation,

[7] C. E. Moore, *Atomic Energy Levels*, NBS Circ. 647, Vols. I and II (National Bureau of Standards, Washington, D.C., 1949 and 1952).

[8] W. C. Wiese, M. W. Smith, and B. M. Miles, *Atomic Transition Probabilities*, NSRDS-NBS 22, Vols. I and II (National Bureau of Standards, Washington, D.C., 1969).

the absorption of interstellar radiation, and the determination of element abundances. The extension to heavier elements is in progress.

Most of our understanding of molecular structure derives from spectroscopic observations, mainly in the infrared and microwave regions of the spectrum. Molecular spectroscopy gives information on not only the energy states and lifetimes but also the relative locations of the atoms, bond angles, interatomic distances, force constants, and moments of inertia and bond strengths. Molecular data lie at the heart of chemistry and are essential to a variety of practical applications.

In recent years many innovative techniques have been introduced that enhance and extend the use of spectroscopy. The most dramatic are the infrared and optical lasers, which have permitted observations of Raman and Brillouin scattering. The laser as a spectroscopic tool is briefly discussed in Chapter 10.

Beam-foil spectroscopy is another novel technique that has significantly extended spectroscopic methods. This conceptually simple method uses a charged beam, which is accelerated to high energy and passed through a thin film of a solid material. The transmitted ions are raised to excited states of singly and multiply charged ions, and the subsequent radiation is observed downstream of the emerging beam. Beam-foil spectroscopy has led to lifetime measurements of energy levels in multiply ionized systems, with transition wavelengths in the vacuum ultraviolet region of the spectrum, and to the simulation of lines seen in the solar corona so that the responsible element and its stage of ionization can be reliably identified. Beam-foil spectroscopy also has demonstrated the phenomenon of quantum interference in fine-structure levels of H, He, He^+, Ne, and Ne^+ and revealed numerous spectral lines not previously seen. The measurement of lifetimes for given levels for the five members of the beryllium isoelectronic sequence offers a valuable check on theoretical methods for calculating transition probabilities. Finally, beam-foil spectroscopy has led to a revitalization of the use of small accelerators.

Classical spectroscopy (the identification of spectral lines and their related energy levels) is still a major field of activity. Chemists use spectral lines in the identification of atomic species, and astronomers use them in determining the compositions of stellar atmospheres. Recently, spacecraft observations of the sun and stars yielded spectra in the soft x-ray and far-ultraviolet regions of the spectrum. Many lines appear in the spectrum that are as yet unidentified. The identification will require many more laboratory and theoretical studies of the spectroscopy of highly ionized atoms and could involve doubly excited states of atoms. The investigations of doubly excited states, which play an important role in autoionization and dielectronic recombination, are proving to be of great value in developing an understanding of atomic resonances and their consequences.

D. CALCULATION OF ATOMIC AND MOLECULAR PROPERTIES

There is little doubt that the Schrödinger equation would provide an accurate description of the nonrelativistic properties of atomic and molecular systems if it could be solved with sufficient precision. During the past few years, many new procedures have been suggested for constructing approximate but accurate solutions, and, with access to fast computing machines, a major advance in precision and generality is likely. Computational methods should lead increasingly to numerical experiments comparable in scope to laboratory experiments.

At the heart of the theoretical description of atomic structure is the central-field, self-consistent, independent-particle model. The language of atomic spectroscopy is based on a combination of the independent-particle model and the exclusion principle, which together provide a satisfying conceptual interpretation of the periodic table of the elements and of chemical structures. The early atomic calculations using the self-consistent approximation led to a prediction of x-ray scattering form factors and helped to initiate the field of x-ray crystallography, which is responsible for much of what we know about molecular geometries and the nature of solids. (See the discussion of the relation of atomic and molecular physics to solid-state physics, Chapter 8, Section B.) The central-field model also led to an understanding of the periodic table of the elements and became the underlying concept in descriptions of chemical structure.

Two different models of molecular structure were derived from the central-field model of atoms. In one of these, the molecule is regarded as being constructed from interacting atoms or ions that do not differ greatly from free atoms, as approximated by the central-field model. This valence-bond model, devised by L. Pauling, was responsible for the first adequate picture of the chemical bond. It led quickly and easily from the atomic structure calculations to a good semiempirical theory of the number, direction, and length of bonds in a molecule. This made possible a unification and systematization of descriptive inorganic and organic chemistry. For the first time, molecules were understood, and new molecules could be predicted from a qualitatively reliable and easily visualized model. This perturbed-atom view also led to the crystal field description of the spectra of atoms in a crystal and the ligand field description of Weiner complexes. In the relatively short period of 30 years, this model has become so ingrained in chemical thought that it is taught as fact in freshman chemistry and taken for granted in advanced graduate courses. It is still used as an empirical tool for predicting and understanding new compounds.

The second model of molecular structure derived from the central-field model of atoms is the self-consistent-field or molecular-orbital model of molecules. This model, devised by R. S. Mulliken, is the mathematical generalization of the Hartree-Fock theory of atoms. It, of course, has its origin in the advent of the large computer. Although valence and molecular geometry are more difficult to understand with this model, it has led to systematization and prediction of molecular spectra. Most descriptions of molecular spectroscopy rely completely on this model. A large branch of organic chemistry still is influenced by a semiempirical version of this model, which has allowed a systematic prediction of bond lengths, chemical reactivity, ionization potentials, nuclear magnetic resonance and electron spin resonance spectra, acidity, and reaction mechanisms of planar aromatic organic compounds.

Important side effects were the training of a large group of chemists in the methods of scientific computing and the development of new and significant mathematical techniques. Several of the chemists trained in this area are now directing computer centers devoted to scientific computing. Also, because of renewed interest in this field among physicists, some quantum chemists have become physicists.

Another side effect of this search for new methods is that very accurate results for systems of as many as four electrons have become available. For example, x-ray scattering factors for bonded hydrogen atoms are available that make it possible to locate hydrogen atoms from x-ray crystal data.

The search for improved methods continues, for major problems have yet to be solved. The electronic structure of polyatomic molecules has hardly been touched because of the difficulty of evaluating certain integrals. These integrals are now beginning to yield to analysis. The faster computers becoming available should soon make it possible to do significant calculations on polyatomic molecules. These calculations should lead to new information on potential surfaces for reactions. Since these surfaces are repulsive, they are not easily measured, but kineticists need such information to build models of reaction mechanisms.

The theory of time-dependent phenomena has only just begun. Problems involving the motion of atoms, and molecules in non-stationary states (i.e., collisions) are discussed in the following Section.

The impact of these semiempirical models on science, technology, and industry cannot be overestimated; they have been incorporated into studies of thermodynamic properties, catalysis, and reaction mechanisms and have influenced the training and thought of a generation of scientists and engineers. Modern metallurgy and ceramics also depend heavily on them. Semiempirical calculations with these models are now being incorporated into theories of biological structure, photosynthesis, and the nature of life.

E. ATOMIC COLLISIONS

The central role of atomic collisions in various fields of science and technology is treated in Chapter 9. Here we discuss the importance of atomic collisions to basic physics, the intellectual contribution of this field to contemporary physics research.

Basically, a collision between two atomic systems (or between an atom and an elementary particle such as an electron, positron, or proton) can be described by the Schrödinger equation or its relativistic counterpart, although the interaction is much more involved than that which takes place within a single atom. The dynamic nature of this problem generally requires a time-dependent description. A collision by its nature describes the temporal evolution of the state of two interacting systems; therefore, collision studies differ from those of isolated atoms and molecules, for which a time-independent description usually is adequate. The computational ability of the physicist is challenged to the utmost by the demands of this problem. Only the simplest of collision problems can be computed without the use of gross approximations. A close and continuing interplay of theory and experiment are essential for progress in this field. Its appeal for the quantum many-body physicist lies in this intimate and mutually stimulating relationship between theory and experiment. Each advance in one stimulates new and more sophisticated approaches in the other. Some examples of the importance of electron–atom collisions follow.

Studies of electron–atom collisions provided much of the early knowledge of the structure of atoms and molecules. In recent years, such collision work has been the most fruitful testing ground of collision theory. Among the more dramatic contributions was the discovery of resonances by G. J. Schulz, attributable to compound structure effects in electron–atom and electron–molecular scattering. These compound states have their analogues in other branches of physics. However, in atomic and molecular physics, it is possible to calculate resonances with much greater computational sophistication than usually is possible in other fields. Further, experimental studies of electron resonances in atoms have led to an essentially new method of performing atomic structure studies, since these resonances are associated with specific states of the compound (electron + atom) system.

The development of high-energy-resolution electron guns has been the significant technological advance that made resonance studies possible. Such devices are also applicable to energy-loss experiments, which contribute an essentially new and very rich complement to ordinary spectroscopy. Electron energy-loss spectroscopy, like ordinary optical spectroscopy, gives the energy levels of excited atoms relative to the ground state. Such spectrographic studies with monoenergetic electrons possess resolution comparable with and possibly soon to exceed, that of optical spectroscopy and have the additional advantage of permitting energy determinations of states that do not emit light.

Compound states exist not only in atoms but also in molecules. In the latter, the compound state, once formed, can emit an electron, leaving the molecule in various vibrational states. Alternatively, a negative ion plus a neutral fragment may result. The cross section for the excitation of vibrational levels via compound states is very large. In fact, a molecular gas laser, such as the CO_2 laser, would not be nearly so efficient if the compound state did not provide the vibrational excitation that results in population inversion.

The postulate of the compound state supplied the first real understanding of negative ion formation. For example, there have been recent and realistic calculations for dissociative attachment and three-body attachment in molecules such as H_2 and O_2. A wealth of phenomena has been discovered in compound-state studies.

Soon it may be possible to conduct experiments using polarized beams of electrons and atoms. A great increase in the knowledge of atomic and molecular structure and understanding of the effects of the exclusion principle—one of the basic postulates of quantum mechanics—can be anticipated as a result of this development.

Experiments in which essentially monoenergetic fast electrons are scattered by target gases are complemented by multiple collision or swarm experiments, which involve the study of the passage of charged particles through gases. Such experiments yield transport coefficients, such as diffusion, mobility, and attachment coefficients, or relaxation coefficients, such as temperature decay times and recombination coefficients. These transport and relaxation coefficients characterize the interaction of electrons, ions, and molecules under gas-kinetic conditions and, as such, are of direct interest to the aerodynamicist, plasma physicist, student of the stellar atmosphere and interstellar space, fluid-mechanics physicist, and others.

The transport and relaxation coefficients are related to the cross sections of single collision experiments through averages over scattering angle and over the appropriate distribution of relative energies of collision. The measurements are analyzed by solving the Boltzmann equation. They provide insight into the implications of the theory of the Boltzmann equation, which governs the motion of electrons in gases and plasmas; therefore, they are of great value in effecting a comparison of theory and experiment in low-density gas dynamics and plasma physics.

Electron scattering studies are of great technological value, and the range of applications is extensive. These aspects are discussed in Chapter 11.

Compound states occur also in atom–atom and atom–molecule collisions. Undoubtedly, such states play a significant role in chemical reactions; much fruitful research in this area lies ahead.

In general, the study of heavy-particle collisions constitutes an interface between physics and basic chemistry in which research activities are broad and diverse. Such reactions are discussed in greater detail in Chapter 9, Section A, in which we consider the relation of atomic and molecular physics to chemistry.

8 Relation of Atomic and Molecular Physics to Other Branches of Physics

The unity of physics is illustrated by the close connection among its various branches. The same relatively small set of basic laws of nature governs the behavior of all physical systems, be they photons, isolated atoms or nuclei, pairs of interacting particles, or conglomerates in various states of matter. Therefore, it is not surprising that atomic physics should be closely connected to its sister disciplines and that such interconnections should occur on many levels, from the most fundamental through the most technologically applied.

The purpose of this chapter is to illustrate the closeness of physics subfields by discussing the relationship of atomic physics to solid-state physics, plasma physics, astrophysics, and nuclear and high-energy physics. This discussion is necessarily selective; it is intended to show, first, the central role atomic physics plays in contemporary physics research and, second, the interrelation and interdependence of all branches of physics.

A. NUCLEAR PHYSICS

The close relationship between atomic physics and nuclear physics predates the time when nuclear physics became a separate discipline. The reason is simple. Since every atom contains a nucleus, nearly all atomic measurements are affected by the electron-nuclear interaction.

The most prominent nuclear feature, from the point of view of atomic physics, is the nuclear charge. Historically, the uniform progression of nuclear charge through the periodic table was first confirmed by an atomic effect, namely, the characteristic x-ray series. An equally fundamental property of nuclei, their intrinsic angular momentum, was first postulated by W. Pauli in 1924 to explain regularities observed in another atomic effect, in this case optical hyperfine structure. An experiment by F. Rasetti used a feature of molecular spectra to prove that the nucleus does not contain electrons. Optical and x-ray spectroscopy helped to establish *K*-capture as a real nuclear phenomenon. Currently, atomic spectroscopic studies are of increasing interest in nuclear physics because they can detect many details of nuclear structure, such as the actual distribution of the charge and magnetism within the nucleus and the variation of these quantities with increasing atomic charge and atomic weight. The systematic measurement of nuclear spins by atomic spectroscopists provided much of the basic information needed by M. G. Mayer and J. H. D. Jensen for the formulation of the nuclear shell model that contributes much to our present-day understanding of nuclear properties. This model is analogous to that provided for atoms by the periodic chart of the elements.

Measurements of atomic hyperfine-structure data permit the determination of magnetic dipole and electric quadrupole moments of nuclei. The electric quadrupole moments are a measure of the general shape of the nuclear charge—they can show whether nuclei are spherical or ellipsoidal, a crucial point in connection with the fission of uranium. The results indicate that the shell model is not an adequate description for all nuclei. The large distortions from spherical shape, combined with information obtained by purely nuclear techniques, thus served as a basis for the collective model first formulated by A. Bohr and B. Mottelson.

Atomic experiments are not restricted to the measurement of nuclear spins and moments. As in the case of high-resolution optical spectroscopy, details of the charge distribution and intrinsic magnetism within the nucleus also have been profitably investigated. In regard to the nuclear charge distribution, the quantity obtained is the change in the charge radius as more neutrons are added to form the isotopes of a given element. The experimental measurement is that of the isotope shift in spectral lines of the medium-weight to heavy nuclei. The making of such measurements dates from the same period as the development of hyperfine-structure measurements. The interpretation in terms of nuclear effects is more recent, dating from about 1930. These measurements compare favorably with those recently obtained in μ-mesic atom studies. (The latter studies, incidentally, also have a large atomic component, since they study essentially the spectra of "hydrogenic" atoms in which the electron is replaced by a negative muon.)

Recent advances have permitted the study of optical spectra of radioactive isotopes with samples of less than 10^{12} atoms, or about 1 ng! Plans for future developments include the use of high-resolution image intensifiers to improve the sensitivity of these investigations. New technological applications of such sensitive detection techniques also are likely.

Recent advances in the atomic beams magnetic-resonance method have made possible the measurement of isotope shifts of radioactive atoms to which the beams technique has already been applied. This is a rather large class of nuclei for which other measurements often are not available. The μ-mesic atomic method is essentially restricted to stable or very-long-lived nuclei that are available in sizable samples.

In lighter nuclei the isotope shift is mass-dependent. The best known example of an exploitation of this characteristic is the discovery of deuterium. In studies of nuclear magnetism, atomic physics methods have been used to determine the nuclear hyperfine-structure anomaly, which is sensitive to the distribution of nuclear magnetism and thus constitutes a unique test of nuclear wavefunctions. To date nearly all values of the hyperfine anomaly have resulted from high-resolution atomic experiments.

Atomic and nuclear physics have progressed in a parallel fashion in the measurement of nuclear masses. Historically, mass spectroscopy preceded the study of nuclear reaction Q values. As nuclear technology developed, the mass difference obtained from Q values became increasingly precise, until today the mass differences that they yield equal or exceed in precision those obtained from mass spectroscopic measurements. The "competition" has afforded fruitful and mutually beneficial interaction between nuclear and atomic physics.

The production of polarized electrons currently is receiving increasing attention. Here the nuclear applications are related to high-energy electron scattering, particularly inelastic scattering from nuclei. In the past few years, several atomic techniques have been developed that will yield beam intensities and polarizations much larger than those obtained by the traditional and technically difficult Mott-scattering method. Some promising examples of these are electron scattering from heavy atoms such as mercury; photoionization of polarized alkali atoms; electrons ejected in the de-excitation of two colliding, oriented, metastable helium atoms; and the collisional ionization of oriented, metastable hydrogen atoms. Many other proposals and experiments have appeared. Probably high-intensity polarized electron beams will become available soon for nuclear research. Of even greater interest in traditional nuclear physics are the atomic techniques for producing oriented ion beams, since many of these are adaptable to use with Van de Graaff accelerators, cyclotrons, and the like. The ions that hold the most interest are the positive and negative ions of hydrogen and helium. In the case of the hydrogenic atoms, the useful methods are nearly all outgrowths of atomic-beam experiments. The atomic hyperfine interaction is used for producing the nuclear polarization and for either deflection in an inhomogeneous magnetic field or selective polarization. For ^{3}He, polarized beams have been produced by optical pumping of He metastable atoms. Other methods, such as one that involves the use of the special properties of an autoionizing state of $^3He^-$, have also been suggested.

The atomic methods enjoy a considerable advantage over direct nuclear polarization techniques, since they work essentially with an electron spin moment 2000 times larger than nuclear moments. Because of this advantage, it is likely that new techniques will rely strongly on the results of atomic physics investigations.

A related problem is the polarization of targets for nuclear scattering experiments. For the important case of ^{3}He, optical pumping has produced a useful polarized target.

Although the nuclear densities of oriented atomic samples are low compared with those needed for nuclear targets, they are ample for studying the decay of oriented nuclei. This type of study, with the orientation produced in solids, has already received considerable attention. Probably atomic methods, which produce high degrees of orientation, will be used more extensively in the future. Interesting examples of one atomic method are the studies of the decay of ^{19}Ne and ^{35}Ar oriented by intercepting particle beams separated by a Stern-Gerlach magnet. These studies yielded information about nuclear β-decay matrix elements and parity-conservation questions.

There are many other important applications of atomic and molecular physics to nuclear physics. Thus, ionization in gases underlies such vital nuclear devices as the cloud chamber, Geiger counter, proportional counter, and spark chamber. Nuclear magnetic resonance is widely used to measure the magnetic fields with which nuclear experiments are concerned. Even the energy level schemes used to describe nuclear systems stem from the prior work on atomic energy levels. On the other hand, nuclear physics has provided reciprocal benefits to atomic physics in, for example, the development of fast electronics, multichannel analyzers, and computer-controlled devices. Thus the two subjects remain closely allied.

B. SOLID-STATE PHYSICS

A close connection between fundamental work in atomic physics and solid-state physics has always existed. Although the flow of ideas has not been unidirectional, frequently concepts and theoretical methods were first developed in connection with atomic systems and later adapted to the solid state. Perhaps one of the most significant examples is the self-consistent field method for the determination of wavefunctions and energy levels for many-electron systems. This method was developed first by D. Hartree and his collaborators in England and by V. Fock in the Soviet Union, soon after the introduction of quantum mechanics about 40 years ago. They recognized that systems with even as few as two electrons are too complex for an exact solution of the Schrödinger equation to be practical, and they introduced the fruitful concept that electrons in atoms could be regarded as independent particles, each of which moves in the field of the atomic nucleus and the average field of all the other electrons in the system. This idea was combined successfully with the requirements of the Pauli exclusion principle and the general variational principle of quantum theory to yield the Hartree-Fock equations. These are a set of nonlinear integro-differential equations for the wavefunctions of the atomic states. This procedure has been extended to electron–atom scattering in the close-coupling method.

Because the equations are quite complex, solutions must be obtained numerically. In the 1930's, when only desk calculators were available, the process of calculating an atomic self-consistent field was lengthy and results were obtained for relatively few elements. In recent years, the availability of large computers has made calculations both more rapid and more precise. Wavefunctions are now available for a large number of atoms and ions. However, substantial gaps still remain, particularly for heavy elements in which relativistic effects are significant. It is still not possible to find in the literature wavefunctions for all the atomic states that are of interest in other physics subfields, particularly solid-state physics.

The applications of self-consistent field methods to problems in solid-state physics were of two types. First, physicists soon realized that the Hartree-Fock equations could be generalized easily to solids and that they offered a satisfactory formulation, in principle, for the calculation of the electronic structure and properties of solids. With suitable generalizations, the Hartree-Fock method is applied to problems, such as the theory of superconductivity, that are far removed from the original applications; much present research is concerned with attempts to find further generalizations and to remove some of the restrictions inherent in the procedure.

Second, the numerical results of specific atomic self-consistent field calculations are of great importance in solid-state physics. One outstanding example is energy-band theory, which attempts to calculate the electronic properties of solids beginning with a knowledge of crystal structure and of wavefunctions for individual atoms. This vital area of solid-state physics would have remained closed to serious investigation if atomic wavefunctions had not been available. Improvements in the accuracy and coverage of such calculations can be expected to yield corresponding improvements and extensions of results in the theory of solids.

Three examples of the manner in which attempts to improve the self-consistent field technique have led to applications in atomic, molecular, and solid-state physics are the following:

The Quantum Defect Method Self-consistent field calculations do not produce energy levels in perfect agreement with experiment. Therefore, the possibility of using observed experimental information from atomic spectroscopy directly, where possible, to determine valence electron wavefunctions at large distances from the core of closed shells, where the potential is known to be coulombic, was suggested. Procedures for doing so were developed, with applications in solid state in mind, by J. H. van Vleck, H. Brooks, and their students. Calculations of cohesive energies, lattice constants, compressibilities, and nuclear magnetic resonance frequencies (Knight shifts) were made for the alkali metals. Subsequently, returning to atomic physics, the method was extended, by M. J. Seaton and collaborators, to the calculation of excitation and ionization cross sections for reactions involving atoms and ions of interest in astrophysics and atmospheric physics.

The Pseudopotential Many of the complications of the self-consistent field procedure can be avoided by the introduction of an artificial potential whose parameters are determined from experimental data. This potential can be so chosen that the lowest state in the field is the ground state of a valence electron in the atom; the effects of electrons in closed shells are replaced by a repulsive term in the potential. This procedure was introduced by H. Hellmann in the Soviet Union in 1935. Hellmann first considered molecules involving alkali metal atoms (K_2 and KH) but subsequently studied the cohesive energies of the alkali metals. The method remained dormant for many years, probably because of lack of adequate computational machinery to determine accurately the parameters of the potential. In the past ten years, the method, now called the pseudopotential, has experienced a revival and presently is quite popular in solid-state physics, where it forms the basis for one of the most widely used methods of studying energy bands. It is now possible to extend such approaches to the calculation of such diverse properties of solids as elastic constants, lattice vibration spectra, and defect energy levels. The time is ripe for a reintroduction of the pseudopotential into atomic and molecular physics, in which it could lead to substantial simplifications in the study of atomic interactions.

Polarization The fundamental reason that self-consistent field calculations of atomic energy levels disagree to some extent with experiment is that, in this method, the detailed interaction or correlation between electrons is neglected. In some significant cases, particularly when one electron is found outside a closed shell, matters can be improved by considering polarization. The distortion of an atom or atomic core by an external charge can be calculated. This distortion, which in some ways is analogous to ocean tides produced by the moon and the sun, changes the field produced by the other electrons, and the change reacts on the external particle. During the early 1930's, polarization effects were considered by J. Holtsmark in the scattering of electrons by atoms and by H. Bethe in regard to atomic structure. About ten years ago, physicists realized that such effects would be important in solid-state physics and, in particular, could account for some of the discrepancy still remaining between theory and experiment in regard to the cohesive energies of the alkali metals. The effect of atomic polarization in other solid-state problems, in which it is intimately connected with the study of many-body effects in general, awaits investigation.

Atomic polarization also enters into optical model calculations of the cross section for the scattering of electrons by atoms. Recent improvements in the theory should find application to the study of energy bands in solids.

We have emphasized so far only one particular area in which there has been a strong mutual interaction between atomic and solid-state physics. Many other areas of equal importance exist and deserve at least brief mention, although limitations of space do not permit a detailed description. One is the theory of the shape of spectral lines emitted by atoms in a medium containing charged particles; this theory has developed in parallel in atomic physics, with reference to plasmas and astrophysics, and in solids, in which particular problems arise in regard to excitons in insulating crystals. Another area, the current importance of which is difficult to overemphasize, is the theory of lasers. The study and further development of both gas and solid-state lasers involves the solution of common problems in excitation transfer and the theory of coherent radiation. Additional areas of common interest are those involving the direct interaction of individual atoms and molecules with solids at surfaces; for example, sputtering, crystal growth from the vapor phase, insulating-to-metallic transition in dense metal vapors, and penetration of atoms into solids.

A particularly interesting and practical application of atomic physics to the solid state is the use of high-energy particle beams to study solid materials. The discovery of fission and the development of nuclear reactors directed the attention of materials scientists to the effects of irradiation on the properties of solids. A thorough understanding of these effects requires detailed knowledge of the way in which energetic atoms and ions move in solids, these being either foreign atoms, such as fission fragments and particles from an accelerator, or atoms of the solid itself, displaced from their proper positions by collisions with fast neutrons or other particles. At sufficiently high energies, the interactions of heavy particles with solids do not differ greatly from what they would be with a dense gas or, in the case of conductors, with a dense plasma. An energetic ion entering a solid may lose its energy through elastic collisions with lattice atoms, inelastic collisions with bound electrons through ionization and excitation, or inelastic collisions with conduction electrons through the generation of plasmons. Lattice atoms liberated by these collisions can cause secondary collision cascades, which may result in the generation of permanent lattice defects (radiation damage) or in the ejection of atoms from the solid surface (sputterings). Manifestations of the electronic interaction can be observed in such processes as the ejection of secondary electrons, the formation of electron–hole pairs in semiconductors, and the change of the ionic charge of an energetic projectile in the target.

Hence, a detailed knowledge of interatomic potentials, ionization and excitation cross sections, and other parameters available from atomic collision studies is invaluable in assessing effects occurring in solids; conversely, information obtained from solid-state collision studies improves the understanding of single-collision processes. This cross-feeding of information is perhaps best illustrated by a newly developing field, ion crystallography, which has evolved rapidly from theoretical calculations on radiation damage using realistic interatomic potentials.

These calculations show that atoms entering the target crystal in certain directions penetrate exceptionally deeply into it. The directions of greatest penetration were associated with channels in the target crystal, that is to say, with regions of the crystal symmetrically surrounded by several long, straight rows of closely packed atoms. A projectile is constrained to move in a channel by making a large number of glancing collisions with the rows of atoms constituting the walls of the channel. In each collision, the moving atom is deflected only slightly, the cooperative effect of many collisions with adjoining target atoms serving to steer the projectile away from the row of atoms and back toward the channel axis. Since only a very small energy loss occurs at each collision, the channeled projectile is able to penetrate much more deeply into the target than an unchanneled particle. This phenomenon, now termed axial channeling, was quickly confirmed experimentally, thus initiating the study of crystals by means of energetic ion beams–ion crystallography.

The largest and most direct application of atomic collisional information to solids is in semiconductor radiation detectors that perform as solid-state ionization chambers. Related to this application is the rapidly developing field of ion implantation. Another direct application is the study of sputtering phenomena, that is, the interaction of ion beams and surfaces (discussed in greater detail in Chapter 11).

Just beginning to be exploited is the use of atomic beams as a probe to study solids and fluids. For example, experiments are being performed in which state-selected (polarized) atoms are "flopped," that is, caused to change their polarization, as they pass by the magnetic vortex structure of a type II superconductor. The atoms serve as miniature moving magnetometers that can probe a wide variety of systems. In another application of atomic physics to cryogenic systems, the velocity distribution of helium atoms boiling off the surface of liquid helium is being studied. Useful information concerning quantum fluids is obtained in this way.

The measurement of magnetic fields is crucial in many solid-state experiments. In addition to the usual methods, a new technique based on optical pumping has been developed recently. One version of this technique has been developed into a highly sensitive mercury detector, while another version is suitable for the detection of weapons. These represent unusually important practical applications of basic atomic physics.

C. PLASMA PHYSICS

Plasma physics can be described as the study of fluid systems in the ionized state. Although the plasma state has not been extensively studied and is not thoroughly understood, it is a rapidly growing, enormously active field, with ever-increasing numbers of scientists attacking both fundamental and applied problems. The 1970 Nobel Prize awarded to H. Alfvén denotes recognition of the place plasma physics now occupies in science. Among the chief motivating factors for the recent laboratory research in plasma physics are the search for controlled thermonuclear fusion (CTR) and the rapid growth of astrophysics. (The relation of atomic physics to CTR is discussed in Chapter 11, Section A. Astrophysics is discussed in Chapter 8, Section E, and the use of hot plasmas in atomic spectroscopy studies, one of the most important applications of plasma physics, is discussed in Chapter 8, Section D.)

Plasma physics bears perhaps the closest relation to atomic physics of any of the physics subfields, particularly the study of low-density plasmas in the gaseous state, in which research activity is greatest. The laboratory formation of the plasma represents a complex problem in fundamental atomic physics processes. For example, the classic method for producing a laboratory plasma is the electric discharge. A discharge can be described as a change of phase from the neutral to the ionized state of a gas. An understanding of this phase change, that is to say, the ability to describe the temporal change in the gas as it ionizes, depends on knowledge of atomic collision and radiation processes. The various electron–molecule elastic, inelastic, and ionization cross sections that govern the breakdown process must be known. Thus the central role of atomic physics in understanding plasmas is immediately apparent.

When the plasma has reached its quiescent state, which may last for only a few microseconds in some high-energy plasmas or for many hours in, for example, the low-density glow discharge of the high-pressure mercury arc, atomic physics again provides the key to understanding. Such a nonequilibrium, steady-state system is governed by the balance between the loss and creation of charged particles. Apart from stability considerations, the list of loss and gain processes is essentially once again a list of electronic and atomic collision cross sections of various kinds. Often many dozens of such reactions must be considered to achieve a proper description of the steady-state discharge.

An interesting application of atomic physics to the generation of high-temperature, that is, thermonuclear, plasmas is in the injection problem. This is a method of generating a hot plasma by firing a high-energy beam, produced externally, into the plasma volume. The charge-to-mass ratio of the beam is increased by some sort of breakup process in the plasma volume, enabling the plasma magnetic field to trap the energetic particles and acquire local equilibrium by collisions with previously produced plasma particles. This breakup can be produced by collisions or by field ionization. An example of the latter type of process is the injection of a neutral beam of atomic hydrogen, in which the atoms happen to be in high-lying excited states. For such atoms, the additional energy required to ionize is far less than for those in the ground state. When these fast-moving, excited atoms reach the plasma, the confining magnetic field appears as a motional (Lorentzian) electric field. This electric field seen by the moving atom may be strong enough to remove the valence electron from the atom, thereby leaving behind an ion that is immediately trapped in the plasma. This fast ion also serves as an energy source whose energy is rapidly dispersed through the plasma by elastic collisions. This field ionization of an excited atom demonstrates the application of quantum theory to a simple atomic system; it illustrates the direct use of atomic physics theory in the design of an effective high-energy plasma injection device.

Another application of atomic physics to plasma generation is the method of ion-cyclotron heating. This is a particularly effective means of energy transfer in a plasma, since the heavy ion is very efficient in transferring kinetic energy to plasma particles by elastic collisions. The corresponding energy transfer by high-energy electrons, for example, is a far less efficient process because of the small electron mass.

Still other applications involve the thermal generation of plasmas, usually employing alkali metal vapors. Here the low ionization energy of the alkali metal enables one to produce ions merely by heating a surface substrate; the valence electron of the alkali is lost to the heated metal, which is so selected as to have a work function larger than the ionization energy of the neutral atom. This effective ionization method, determined primarily by the peculiar atomic structure of the alkalis, results in a particularly useful type of plasma. Such a plasma is extremely quiescent, since no violent process, such as a discharge, is used to produce it. An entire class of quiescent plasmas, dominated by the so-called "*Q*-machine," has been built as a result of this surface ionization phenomenon. A still more placid plasma in a state of complete thermodynamic equilibrium can be produced by enclosing the alkali vapor in a hot container. Such equilibrium, or Saha, plasmas are in an early stage of development.

For all plasmas, energy-loss mechanisms to the plasma container, either by particle transport or by radiation, are obviously of central interest to the plasma physicist. He often would like to maintain the plasma for the longest possible time and with the minimum expenditure of external energy. Atom–ion charge transfer, bremsstrahlung (radiation from scattered or magnetically trapped electrons), elastic collisions, and radiative decay are examples of atomic physics events that contribute to the difficult problem of plasma losses and the calculation of these losses. Only by thorough exploration of each of these effects can the loss processes be understood and controlled.

Atomic and molecular phenomena play an equally important part in plasma diagnostics. Electron temperatures are determined from spectral line measurements, laser beam scattering, and bremsstrahlung measurements. To interpret the spectral line intensities one has to know the oscillator strengths for the transitions. The measurement of ion temperatures relies to a large extent on measuring Doppler shifts and broadening. Ion energy distributions, which are of utmost importance in plasma dynamics, are usually determined by measuring the flux and energy distribution of energetic neutral particles that escape from the plasma region. This distribution must be unfolded; to do so requires detailed knowledge of the efficiency of neutral detection and the electron capture cross sections of plasma ions in residual vacuum gases. The composition of the residual gases must be known. Ion and electron densities are usually determined by neutral particle detection, spectral line broadening and absorption, merging of line series into the continuum, laser beam scattering, and absolute line intensities. Of increasing importance is the projection of an electron, ion, or neutral beam of particles across a plasma to determine densities and plasma potentials.

D. PLASMA DIAGNOSTICS–SPECTROSCOPY

The complex and frequently hostile plasma environment represents one of the physicist's greatest challenges. All of his traditional diagnostic tricks experience special difficulties when used to investigate the plasma state. The most fruitful diagnostic tool, as in many other branches of physics, is spectroscopy. The need to improve our understanding of the radiation problem, as applied to plasmas, has caused a resurgence of quantitative (or plasma) spectroscopy. Plasma spectroscopy can be defined as the measurement and interpretation of radiation from high-temperature gases. It provides not only diagnostics for laboratory plasmas but also enables the astronomer to understand stellar atmospheres, including that of the sun.

As an example, the wavelength profile of a particular line usually contains information concerning the density and temperature of species (such as electrons) in the immediate vicinity of the emitter. The understanding of line shapes requires a combination of (a) statistical mechanics and many-body theory and (b) atomic physics and collisional cross sections for excited states of the atom. Thus, the interpretation of Stark and Van der Waals broadening requires the blending of many related disciplines.

Since many laboratory plasmas are transient, with time scales of the order of microseconds, it has been necessary to develop instruments suitable for the rapid measurement of line profiles of less than 1 Å width. This has been successful and has led to the construction of fast-response, high-resolution image tubes.

The recent effort with respect to oscillator strengths, discussed in Chapter 6, Section D, has been of direct benefit to the plasma-diagnostics problem. As our understanding of the plasma state improves it is likely that the interplay between pure plasma aspects and atomic physics aspects of the plasma problems will grow. There will be an ever-increasing feedback between the two disciplines, to their mutual benefit.

E. ASTROPHYSICS

Atomic and molecular physics has always been important to astrophysics, but its critical role in the interpretation of astronomical phenomena has perhaps only become apparent recently. The increasing sophistication of astronomical observing techniques is opening new regions of the electromagnetic spectrum in which the universe can be observed; the interpretation of the observations requires greatly improved precision in the basic data of atomic and molecular physics. Not only is greater precision required but so also is a deeper understanding of the nature of the extraordinary variety of atomic and molecular processes that can occur in the diverse physical circumstances obtaining in the universe. For example, in the solar corona, which is the source of most of the radiation producing ionization in our atmosphere, electron densities in active regions can be of the order of 10^{13} cm^{-3} and temperatures of the order of 10^7 K; in the interstellar clouds, from which stars are formed, the particle densities are of the order of 10 cm^{-3} and temperatures are of the order of 50 K.

As an example of the close connection of atomic and molecular physics to astrophysics, consider first the observation of the spectrum of a star. From it we wish to deduce the nature of the star, its composition and temperature, its sources of energy, the conditions of its interior, and its correct place in its evolutionary path. We attempt to do so by acquiring a quantitative understanding of the atomic and molecular processes by which some of the internal energy emerges as the observed luminosity, by assuming such parameters as element abundances, energy fluxes, and gravity, and by constructing a model stellar atmosphere, from which the emergent spectrum can be calculated. A comparison of the model predictions with observation initiates an iterative process in which the model is revised to achieve a close agreement with observation.

To limit the number of possible models, it is desirable to observe a very large number of spectral features such as the strength and nature of the radiative continuum and the shapes and strengths of many different spectral lines. To work out the model predictions requires substantial amounts of computer time and large quantities of basic atomic data. Major uncertainties attend these data. In addition, existing data are scarce and generally are not sufficiently accurate. There is an important need for the development of more reliable radiative and collision cross section determinations, both experimental and theoretical. Our ability to understand the dynamics and constitution of stellar atmospheres depends on our detailed understanding of the atomic and molecular processes that produce the observable radiation.

Our limited information about the remarkable quasi-stellar objects also comes from attempts to interpret the observed spectrum by using atomic data.

In the outer reaches of a stellar atmosphere the corona occurs. It is a region of high temperatures of the order of millions of degrees in which many normal species appear stripped of many electrons. In the solar corona, for example, Ne IX, obtained by removing eight electrons from the gas, neon, is a prominent constituent. It has been suggested recently that Fe XXV, obtained by removing 24 electrons from iron, is an important contributor to the x-ray flux observed from the astronomical object Scorpius X-1. Both Ne IX and Fe XXV are isoelectronic with helium. In such hot, highly ionized circumstances, a variety of unusual atomic processes become significant, not all of which are clearly understood. One is dielectronic recombination, which may be the dominant plasma-loss mechanism in a coronal region. It occurs when a free electron excites a positive ion, forming a temporarily excited state, which subsequently stabilizes by emitting a photon. Much experimental and theoretical work is required for quantitative understanding of this mechanism, which occurs generally in circumstances of high temperature and high fractional ionization.

The observations, by rocket-borne instrumentation, of radiation from the helium-like ions, such as Ne IX, in the solar corona have recently stimulated important advances in our knowledge of conditions in the solar corona as well as of the basic interaction between matter and radiation as formulated in relativistic quantum electrodynamics. It was generally accepted that the metastable 2^3S states of helium-like ions decayed by a two-photon emission:

$$X(2^3S) \rightarrow X(1^1S) + h\nu_1 + h\nu_2 .$$

Consequently, the metastable states were very long-lived. Recently, however, the emission lines of many helium-like ions, produced by the single photon decay,

$$X(2^3S) \rightarrow X(1^1S) + h\nu,$$

have been identified in the coronal spectrum. The single photon decay probability is zero in nonrelativistic quantum mechanics, and the emissions are the result of a higher-order relativistic effect, which has now been calculated to high accuracy. The solar observations and the theoretical interpretations led to a series of skillful measurements using beam-foil techniques. A small discrepancy remains between theory and measurement over the lifetime of Ar XVII (2^3S), which might be due to inadequate knowledge of the quantum electrodynamics of electron–electron interactions. In any event, the theory of the decay process and the measurements and calculations of the lifetimes have now been injected into the interpretation of the coronal observations with consequent modifications in the physical description of coronas.

The discovery of the single-photon decay process has important consequences for the interpretation of neutral helium emissions from planetary nebulae and attempts to observe the helium content of the interstellar and intergalactic media.

Nebular emissions are governed by atomic and molecular processes. The temperatures are of the order of 10,000 K, and most of our information about their properties and their evolution and origin comes from the interpretation of emission lines from metastable states of weakly ionized systems and from excited states of hydrogen and helium. The methods for determining electron densities and temperatures of gaseous nebulae depend critically on certain atomic parameters, such as excitation of metastable states.

The vast regions of interstellar space divide into interstellar clouds, where the densities are of the order of 10–100 cm^{-3}, temperatures are of the order of 10–100 K, and fractional ionizations are of the order of 10^{-3}, and the intercloud regime, where the densities are the order of 1 cm^{-3} or less, temperatures may be several thousand degrees, and the fractional ionization, as inferred from pulsar dispersion data, may be as high as 0.2. Interstellar matter plays a critical role in the evolution of the galaxy, serving both as a repository for matter ejected from stars and a reservoir for the formation of new stars. Of special recent interest are some cold dense clouds that have been observed.

The processes by which the interstellar medium is ionized and heated and those by which the clouds cool and thermal instabilities develop that may lead to the formation of new stars are largely within the scope of atomic and molecular physics. Thus, fine-structure transitions of neutral and singly ionized species, caused by collisions with neutral hydrogen atoms, provide a dominant mode of cooling in an interstellar cloud and are probably vital elements in the formation of such a cloud. In some astrophysical objects, fine-structure transitions induced by proton impact at high temperatures are important sources of infrared radiation that must be recognized in the interpretation of infrared observations.

The sources for heating the interstellar gas may be cosmic rays or x rays or ultraviolet radiation from supernovas. In all cases, atomic and molecular processes are the key to the heating that ensues.

Dramatic examples of the close relationship between laboratory radio-frequency spectroscopy and the new observational field of line radio astronomy are the recent detections of complex molecules in the interstellar medium. At the time of writing, radio-frequency observations have detected the hydroxyl radical OH, ammonia NH_3, water vapor H_2O, carbon monoxide CO, cyanogen CN, hydrogen cyanide HCN, cyanoacetylene HC_3N, formic acid HCOOH, and methyl alcohol CH_3OH. It had been known for years from optical absorption studies that CN, CH, and CH^+ are present in the interstellar medium; now it also is known from ultraviolet absorption observations that H_2 is present. To interpret these observations and to suggest further critical tests, greatly increased knowledge of molecular physics is necessary. Such knowledge will come not only from laboratory molecular spectroscopy but also from more comprehensive studies of the processes of molecular formation.

It is instructive to note that the first measurement of what may be the temperature of the blackbody radiation remnant of the "big bang"—the primordial explosion that is believed to be the origin of our present universe—was obtained years ago from the optical observation of the relative absorption by the two lowest rotational levels of CN.

In radioastronomy, recombination lines, formed by capturing electrons in very highly excited states, also are observed. Lines from hydrogen in the level with principal quantum number $n = 254$ have been detected. The atomic physics of an environment in which such diffuse structures can radiate presents interesting problems the solution of which is required for the proper interpretation of the observations.

We mention finally the relevance of atomic and molecular physics to cosmology. If the universe originated in a big bang, there must have occurred an atomic and molecular physics era in which the conditions for the formation of galaxies were created. A large number of atomic reactions would have been necessary.

As was true of other fields discussed in this chapter, progress in astrophysics and atomic and molecular physics must be parallel and concomitant, with each stimulating and complementing the other.

9 Relation of Atomic and Molecular Physics to Other Sciences and Education

Both atomic and molecular physics and chemistry deal with the electronic structures of atomic and molecular systems and the ways in which these are affected by mutual interactions. That any division exists between disciplines is largely the result of divergence of interests rather than fundamentally different physical processes. The same is true of the relationship of atomic and molecular physics to atmospheric science and, to a lesser degree, to biology and health physics.

The systems studied by the different sciences can be enormously complex, as in macromolecules and living organisms. An understanding of simple atomic interactions may be inadequate to the task of constructing, *ab initio*, a fully developed understanding of such systems. Thus, different sciences must develop concepts and techniques that are suited to the objects they study. It is usually difficult to relate these concepts to basic physical principles, and, in some instances, it is not helpful to do so. However, for the purposes of this report, we have chosen those subject areas of other disciplines in which this relationship is most direct. Our discussion deals with applications of atomic and molecular physics to chemistry, atmospheric sciences, biology, and health physics. We also discuss briefly the role of atomic and molecular physics in undergraduate and graduate education.

A. CHEMISTRY

From one point of view, chemistry is but a portion of the physics of atoms and molecules. Thus, essentially all atomic and molecular physics applies in large part to chemistry in its broadest sense. To avoid complete diffuseness in discussing the contributions of atomic and molecular physics to chemistry, we have limited the scope of this section to those parts of atomic and molecular physics that relate to chemical reactions (or rearrangement collisions), although we include some indirect but important influences. Information from atomic and molecular physics on the electronic configurations of atoms and molecules and the geometrical structure of molecules shapes much of a chemist's thinking in regard to chemical reactions. Moreover, to study the reaction, the chemist frequently makes use of such physical techniques as infrared spectroscopy, mass spectroscopy, electron spin resonance, and nuclear magnetic resonance. The laser now makes possible the study of chemical reactions between molecules in precisely arranged initial states of many kinds.

Because of the close relation between chemistry and atomic and molecular physics, it is now becoming commonplace for chemists to contribute many papers to conferences on atomic physics, especially to those concentrating on collision phenomena. The *Journal of Chemical Physics*, published by the American Institute of Physics, is the principal journal for a large number of chemists as well as for many atomic physicists. The Division of Physical Chemistry of the American Chemical Society recently sought and obtained affiliation with the American Institute of Physics. The task of obtaining useful information concerning interatomic and intermolecular forces by experimental and theoretical methods is one in which both physicists and chemists continue to be active. Often only local historical precedent determines the department in which such investigations are conducted. However, application of the methods of atomic physics continues to produce the major advances in these investigations.

In the Authorization Hearing for the National Science Foundation in the Spring of 1969, Philip Handler, Chairman of the National Science Board, said, "Chemistry is undergoing a renaissance. This is a new science. It is, perhaps, oldest of the laboratory sciences but it had fallen into the doldrums about the time of World War II. . . . What has changed chemistry again has been the availability of a new set of tools. . . . Nuclear magnetic resonance, spectrophotometry, very careful spectroscopy, electron diffraction, electron paramagnetic resonance spectrometry, instruments for measuring circular dichroism, mass spectrometry, and a few others. . . ."[9] The apparatus and techniques mentioned here came from advances in atomic physics subsequent to World War II. In the following paragraphs, we briefly consider three aspects of chemistry–structure, collisions, and transport properties.

Structure The experimental determination of both electronic configurations and geometrical structures of atomic and molecular systems is fundamental to all of chemistry. Obviously, theoretical calculations of the electronic states of atoms and molecules rely directly on accurate knowledge of atomic and molecular configurations and molecular bond lengths and angles. Indeed, *ab initio* calculations of molecular electronic structure cannot yet achieve correct theoretical values of bond distances and angles. For very simple systems, recent calculations yield potential energy curves that bear a striking resemblance to the experimental potential energy curves, although the calculated curves are displaced toward higher energies by about a molecular dissociation energy. These calculations represent the best that can be made at the present time. Despite their limited accuracy, they have increased scientists' confidence in their ability to understand the electronic structure of molecules.

The impact of atomic and molecular structure on theoretical chemistry has been fundamental and extensive, from the "Paulingesque" understanding that existed as recently as the early 1950's to much more realistic and sophisticated current understanding. Even the language used in chemistry to describe a variety of processes, ranging from the mechanism of reactions to the setting of ions in crystals, depends on the knowledge obtained from experimental and theoretical research on atomic and molecular structure.

The precise geometrical structures of the majority of the known diatomic and simple polyatomic molecules in their ground states have been established during the past three decades. Much less information has been gathered for molecules in excited electronic or vibrational states. The experimental techniques used have varied widely, including, for example, x rays of solids, electron diffraction of gases, infrared spectroscopy, microwave spectroscopy, and optical

[9]The 1970 National Science Foundation Authorization Hearings before the Subcommittee on Science, Research, and Development of the Committee on Science and Astronautics, U.S. House of Representatives, 91st Congress, 1st Session, HR 4283. Vol. 1 (p. 10). (Dr. Handler's testimony was given on March 17, 1969.)

spectroscopy. Other information such as precise dipole moments, fine-structure splittings, and hyperfine-structure interactions also has been obtained. The case of LiH is a good example of the usefulness of precise experimental data on structure. The wavefunction of this hydride is believed to be reasonably accurate. Into this calculation went the accurate experimental bond distance for LiH in its ground state. The dipole moment, among other parameters, was computed *ab initio* from the wavefunction that was obtained. Agreement between experiment and theory was excellent and offered an outstanding confirmation of the theoretical approach.

Other useful tests of theory are provided by information about fine and hyperfine structure. The parameters that characterize this structure depend on such details of the electron wavefunction as the correlation in the motion of electrons. Little information has been obtained on excited states of molecules, but a number of interesting optical level crossing and double resonance studies have been made on atoms.

Since one of the purposes of atomic and molecular structure determinations is to provide input and test information for quantum-mechanical calculations, we also should mention other relevant experimental techniques. Several different kinds of experimentation give results having a bearing on the disposition of atomic and molecular energy levels. Examples are electron excitation, photoionization, and optical level crossing. These techniques not only produce data on the nature of excited states but also yield interesting and useful information about processes that occur when the experiments are performed. For example, when an electron excitation potential is obtained, cross sections for various electron excitation processes are discovered. The study of electron–helium collisions showed that a transient negative ion state He^- existed; this, incidentally, is essential to the operation of tandem Van de Graaff accelerators with beams of helium. In optical pumping and crossed molecular beam work, evidence for complex formation in simple systems has been found. Recent level-crossing experiments have indicated a transient complex between helium and thallium.

Electron spin resonance has been successfully detected in gases containing species (radicals) with unpaired electrons. When the radicals are atoms, this technique can be used in the study of chemical reactions by observing the change in concentration of the radicals. In the case of diatomic radicals, the spectra show fine structure that yields the rotational constant or bond length, quadrupole coupling constants, and hyperfine-structure constants. For simple molecules, this information is valuable in understanding refined electronic structure calculations. The spectra of polyatomic radicals in the gas phase are more complicated, as past studies of NO_2 have revealed. For these radicals, optical Raman spectroscopy and microwave spectroscopy are apparently more useful. Microwave spectroscopy has provided accurate molecular dimensions for specific states of molecules and thus produced much fundamental information. A great deal of effort here has been directed toward the measurement of barriers to internal rotation in molecules.

Chemical Collisions A proper description of the progress of a chemical reaction as a function of time is a characterization of changes that occur as reactants with specified kinetic energies and internal states. Traditional chemistry studies reactions in bulk and is forced to deal in averages taken over an enormously large number of reacting and product constituents distributed over a broad range of states, with both reaction and inverse reaction going on. Although research of this nature continues to be a bulwark of chemistry, in many chemistry laboratories the atomic-beam-collision techniques developed in atomic physics laboratories during the past decade have been adapted to enable chemists to study individual encounters between pairs of particles in well-defined states and to approach the desired description of chemical reactions.

Much of the progress that chemists have made in understanding what happens during a chemical reaction has come from experiments and theoretical work on simple systems. The simplest possible chemical reaction is the rearrangement collision of three particles, A, B, and C, represented by

$$A_i + (BC)_j \rightarrow (AB)_k + C_1,$$

where the subscripts characterize individual quantum states of the various species. Although these rearrangement collisions are of greatest interest to the chemist, even the simple elastic and inelastic collisions not involving rearrangement provide substantial insight into chemical phenomena. First, the study of elementary elastic collisions supplies the conceptual basis for the description of more complicated collisions. Second, elastic collisions give information on the interactions between atoms and small groups of atoms; that information can be applied to a number of chemically interesting phenomena.

Third, inelastic collisions that involve only changes of internal energy can be considered prototypes of certain chemical reactions.

Elastic collisions give information on the potential energy of interactions. In many cases the interactions between complex molecules can be described as the sum of interactions between small constituent parts of the molecules. Thus a few good measurements on relatively simple systems may be a sufficient basis for the determination of interactions between rather complex molecules of chemical interest. Such information can be applied directly to such chemically interesting areas as the properties of gases at extreme temperatures, vibrational relaxation in molecules, radiation damage, hot-atom chemistry, and the conformational analysis of large molecules.

Recently, collision techniques have received publicity in the popular scientific literature under the newly coined name, "chemical accelerators." These devices are adaptations of the atomic physicists' ion-beam apparatus and make use of the phenomenon of charge transfer, which has been studied extensively by atomic physicists. The particular manifestations of this phenomenon that are important to chemistry are the production and detection of beams of fast neutral atoms and molecules.

Measurements of elastic and inelastic scattering have made it possible in some simple cases to map the changes in potential energy that occur as changes take place in the relative orientations and separations of colliding atoms and ions. An example is the full description now possible of the rate of hydrogen-atom-catalyzed conversion of *ortho*-hydrogen to *para*-hydrogen. Another is the reaction between K and Br_2, for which angular distribution measurements have been made as well as an analysis of the rotational state of the product KBr. The techniques are being extended to reactions involving more complex species.

Although as yet little is known about the relative reactivity of molecules in individual quantum states, chemists believe that various inelastic but nonreactive collisions are an important preliminary to reaction. Thus the study of inelastic collisions is of direct interest. In principle, the simplest method for the study of inelastic scattering is the direct observation of changes in internal quantum states of atoms and molecules as the result of collisions.

When molecules such as I_2 are produced in specific excited states by the absorption of radiation, observation of the fluorescence in the presence of various perturbing gases gives information about the cross section for inelastic transitions of the excited molecule. Alternately, reactive molecules such as $CH_3CH_2CH_2CH_2$ can be formed in excited states as the result of chemical reaction and the rate of loss of internal energy deduced from the nature of the products formed from these chemically unstable species.

Cross sections for inelastic processes such as the transfer of energy between vibrational or rotational states also have been estimated from bulk measurements of relaxation times. Experiments with shock waves and ultrasonic absorption and dispersion often are made for this purpose. Even the most complex reactions in solution usually are regarded as a mechanism, that is to say, a set of isolated elementary reactions or rearrangement collisions occurring sequentially, in parallel, or both. Atomic and molecular physics has made important contributions to chemistry by helping to establish both the nature of the individual steps and the way in which they are combined in the mechanism.

For the study of these individual steps, the simplest method, in principle, is to prepare reactants in known quantum states and examine the cross sections for the formation of products in particular internal states, that is, the reactive scattering. Although much work remains, substantial progress has occurred with the use of molecular-beam techniques. Here, the species involved in the collision are known, so there is no question about the mechanism of the reaction.

Of the reactions studied, some, such as $K + Br_2 \rightarrow KBr + Br$, have large cross sections ($\sim 10^2$ Å^2) and most of the product is forward scattered relative to the velocity of the incident K atom. Others, such as $K + CH_3I \rightarrow KI + CH_3$, have smaller cross sections and the product scatters backward. For still others, such as $K + NaCl \rightarrow KCl + Na$, there is symmetrical scattering forward and backward so that the formation of a collision complex giving at least several rotation times seems to precede the final formation of the products. Polarized beams have been used to study the dependence of the reaction cross section on the orientation of the molecule in an atom–molecule collision.

Analysis of the nonreactive scattering, using the optical model of nuclear physics, provides details of reactive collisions. This analysis yields threshold values for the impact parameter and the potential energy of interaction as well as the energy dependence of the reaction cross section.

For more complex reactions, in which several different types of collision may be involved, the techniques of atomic and molecular physics again are helpful in investigations of the over-all mechanism as well as of each individual type. The rates of depletion of reactants and the formation of intermediates or products have been followed in many gas reactions with the use of electron spin resonance; infrared, visible, and ultraviolet spectroscopy; and mass spectroscopy. Ion-cyclotron resonance, particularly double resonance, is very helpful in the study of mechanisms; it shows which particular molecular ions are coupled by chemical reactions.

A special atomic collision process, the ion–molecule reaction, has been studied for many years by atomic physicists. This kind of reaction, in which an ion colliding with a molecule may exchange with and free one of the molecular constituents, is clearly a primitive chemical reaction. The interests of chemists in such reactions tend to focus on the region of thermal kinetic energies and those of physicists, on a more energetic region, but no clear-cut and rigid division exists. Currently many chemists are turning to the study of these reactions.

Transport Properties In chemistry, the so-called "transport properties"—diffusion, viscosity, and heat conduction—are important. The understanding of these properties and the ability to predict them under new conditions require three fundamental types of information: the general theoretical formalism obtained from statistical mechanics and kinetic theory, a knowledge of intermolecular forces, and information about the rates of exchange of energy among the various internal modes of motion of a molecule. The last two of these are obtained principally from measurements of elastic scattering and the absorption and emission of radiation. In a few cases in which valid comparison has been possible, transport properties have been predicted with a high degree of accuracy.

It is important to understand the partitioning of internal energy among the possible modes of motion of an isolated molecule in its own right as well as for its application to the study of transport properties. Partitioning is a difficult problem on which a concerted attack only recently became possible. The appropriate techniques are largely adapted from atomic and molecular physics. A recent success, reported by Japanese physicists, is the measurement of the rotational diffusion coefficient of tobacco mosaic virus. They developed for this purpose a high-resolution optical heterodyne spectrometer using the 6328-Å He–Ne laser line.

So far, understanding of the liquid phase is incomplete. A knowledge of forces between pairs of particles is unlikely to be sufficient for a description of liquids on the atomic scale. The extension of atomic collision physics to three-body collisions presents formidable experimental problems; no attempt at this extension has been reported as yet. Clearly, such an extension would have far-reaching consequences for understanding not only liquids but also the perplexing problem of phase change.

B. AERONOMY

Atomic physics is so closely linked with planetary aeronomy that progress in this geophysical–astronomical discipline in large measure is dependent on progress in certain phases of atomic physics. We illustrate this interdependence by a discussion of photon, electron, and proton stimulation of the dayglow and the aurora.

The term aeronomy was introduced about 25 years ago to characterize the discipline concerned with the physical, chemical, and electromagnetic processes that take place in the upper atmosphere of the earth and planets. Actually, it is a very old discipline; the study of auroras, an important facet of aeronomy, extends back to earliest historical times.

Rather than attempting to survey all facets of the interrelationship between atomic physics and the atmospheric sciences, we will give only a few illustrations, with special emphasis on an approach to the calculations of natural upper-atmospheric spectral emissions stimulated by photons, electrons, and protons originating from the sun and magnetosphere. These require as inputs a detailed knowledge of many atomic and molecular properties. The results of such calculations are compared with observations made with high-resolution instruments from rocket probes, satellites, planetary flyby vehicles, and the ground.

Spectral Emissions Stimulated by Atomic Particles Figure 5 illustrates a conceptual sequence of steps in the calculation of the upper-atmospheric spectral emissions stimulated directly or indirectly by the sun.

For photons, Step 1 (in Figure 5) requires precise spectroscopic measurement of the extreme-ultraviolet and soft x-ray flux. Such data, just becoming available from satellites, display large temporal variations associated with solar activity. The geometry problem for photons impacting on the earth (Step 2) is relatively simple, with the exception of grazing incidence at sunrise or sunset or in the polar regions. In such cases, instead of a simple secant law for the path

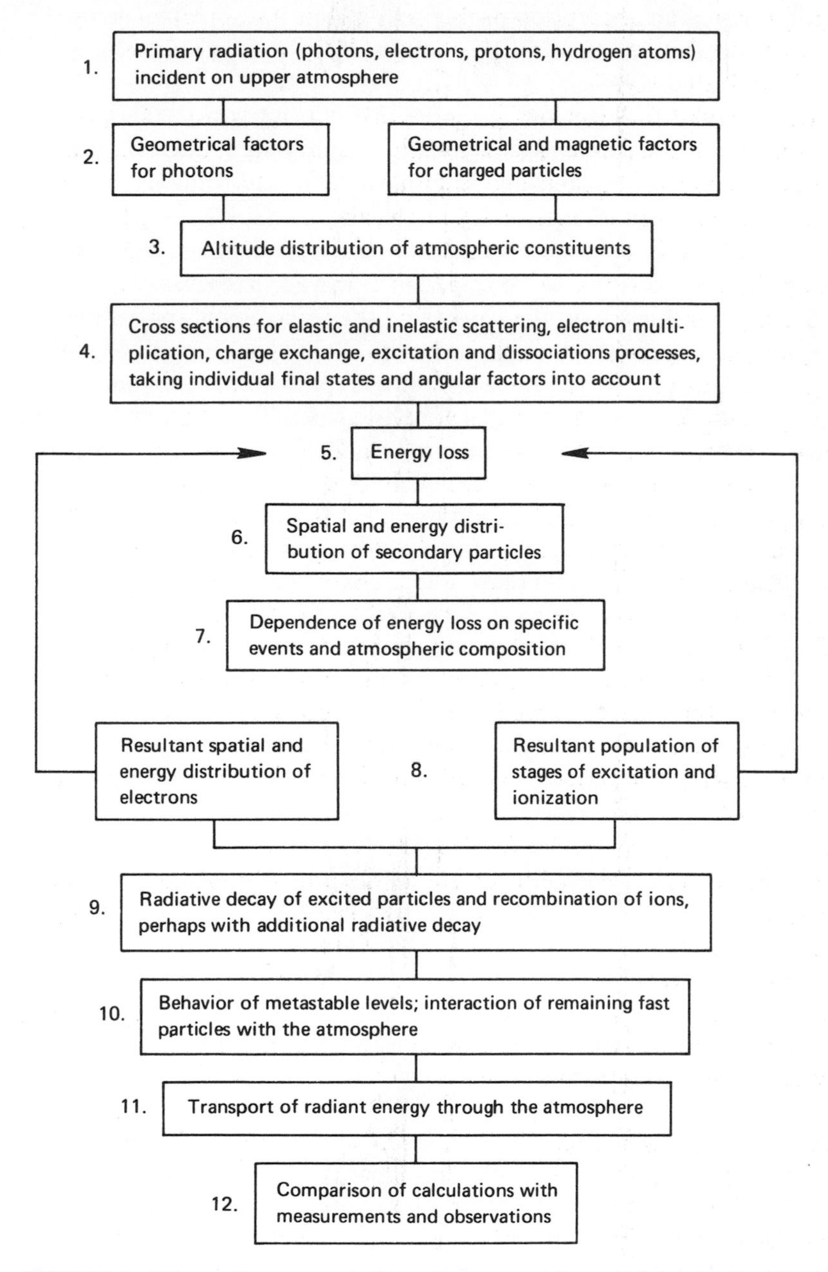

FIGURE 5 Schematic representation of sequence of events in evaluating the effects of radiation on the earth's atmosphere.

length, one has an absorption problem in which the spherical geometry and the altitude distribution of the atmospheric species interact in a complicated way.

Steps 1 and 2 for electrons are more difficult to execute; despite the considerable amount of data already accumulated, electron flux measurements from rockets and satellites constantly bring unexpected results. Very intense fluxes, increasing as the energy goes below 1 keV, have been observed among trapped electrons and also among the electrons that are dumped into the auroral zone. The geomagnetic influence on the passage of the electrons is exceedingly complicated, especially since the magnetic field changes with a geomagnetic disturbance. The same situation applies to protons and hydrogen atoms associated with solar wind, solar flares, and the magnetosphere. Here it should be noted that, unlike electrons, protons can pick up electrons and, as neutral hydrogen atoms, are then decoupled from the magnetic field lines. On stripping in the tenuous upper atmosphere of the earth, these fast-moving particles become recoupled again to the field lines. These charge changes cause a complicated diffusion of protons with respect to the field lines.

Step 3 in all channels involves the specification of the atomic, molecular, and ionic species concentrations in the atmosphere as a function of the altitude. Step 4 for photons involves the totality of inelastic and scattering cross sections that control the ultimate fate of the primary photon. Here one needs cross sections for the following events:

scattering at the angle θ with respect to the direction of the incident photon,
fluorescence,
resonance absorption,
Auger effect,
production of photoelectrons with new energy, and
multiple ionization or excitation.

Step 4 for electrons of energy E depends on the following cross sections:

elastic scattering,
inelastic scattering producing excited state j,
production of secondary electrons with new energies and directions,
dissociation with dissociation energy,

dissociation with release of additional electrons, and
multiple ionization and excitation.

Step 4 for protons requires the complete array of cross sections needed for electrons. In addition, one must consider charge pickup and loss and the interactions of neutral hydrogen atoms with the atmosphere constituents.

The cross sections in Step 4 control the time, spatial distribution, and energy history of primary particles. However, the way to use such cross sections, when they are known, is uncertain. Eventually, a Monte Carlo type calculation might be the most practical way to carry out Step 5. Currently simple techniques for scaling laboratory data are those principally used.

The energy deposition (Step 5) of the primary particles leaves a trail of excited atoms, molecules, and ions and hosts of electrons. These electrons range in energy up to the keV region, but they are predominantly below 100 eV. They have a complex spatial distribution. The photon, electron, and proton problems must be treated in parallel.

These events lead to a new particle distribution (Step 6). Thus, in Step 7, we are concerned again with details of all excitation, ionization, and other inelastic electron-induced processes. Because of the failure of the Bethe-Born approximation and certain sum rules applicable at high energies, detailed characterizations are needed for all inelastic cross sections—not just the specific cross sections related to spectral lines of interest—to keep adequate records on the energy variations with time. Step 8 includes a cycling back to Step 5. However, there is some "leakage," which leads to Step 9. At this level we deal with the fast ($\sim 10^{-8}$ sec) radiative decay of excited states, which requires a knowledge of the entire array of transition probabilities. The slower seconds-to-hours time–scale fate of metastable states and the fate of the active atoms, ions, and free electrons are the concerns at Step 10. Finally, Step 11 deals with the radiative transport of atmospheric emissions and includes the photochemical processes induced by the harder ultraviolet emissions.

The results of these calculations are theoretical numbers that are compared with field observations in Step 12. When preliminary calculations of this type have been carried out, qualitative consistency (agreement to within a factor of 2) between theory and experiment has been realized in many cases. However, when disagreements occur (sometimes they are orders of magnitude in size), the feedback experience indicates that the error is about as likely to be an erroneous atomic property as an erroneous aeronomical property.

Discussion and Conclusion Figure 5 and the accompanying discussion show how closely atomic properties are interwoven with aeronomical problems. The needs of aeronomy focus attention on many areas of atomic physics that are strikingly undeveloped, despite the prevailing view that atomic physics is a highly developed science. There is a particularly large gap between collision theory and experiment on light gases; this has hindered progress in aeronomy, in which collision cross sections are major input parameters. Although the basic interaction and the way to set up the wave equation for many-electron atoms are known, we are far from being able to calculate the cross sections that are needed.

Most of the information about the atmosphere of the planets comes from observations of their emission spectra. A large amount of data about Mars and Venus has been acquired by the Mariner spacecrafts, but the interpretation of the data is fraught with uncertainties, many if not most of which stem from the lack of information about the atomic and molecular processes that occur. Equally serious, uncertainties in the atomic and molecular processes that occur in gases of planetary atmospheres severely hamper the identification of the crucial experiments that should be performed if understanding of the planetary atmospheres is to advance; the space probes are correspondingly much less scientifically effective.

C. BIOLOGY

Until recently, the contributions of physics to biology have usually come about through the application of physics concepts and physics instrumentation rather than through the direct study of biological systems by physicists. Biological and medical research and medical practice are based firmly on physical law and instrumentation, a relationship that is likely to continue. As in chemistry, devices originally developed by physicists for the study of physical problems were applied about a decade later in the biological research laboratory and the physician's office. An obvious example is the x-ray machine. In short, since biological systems are conglomerates of atoms, molecules, and ions, the role of atomic and molecular physics in biology is much the same as it is in chemistry, although of course applications of fundamental concepts of atomic and molecular physics to biological systems are far more difficult than to chemistry.

The connection between physics and biology is becoming institutionalized through the development of biophysics. On a borderline between disciplines, its scope is not yet well defined. It attracts physicists who have a special interest in biology and biologists interested in the more quantitative and analytical aspects of their subject. It has been said that in the nature of the material studied biophysics is much like conventional biology, but in methodology it is closely linked to physics. Our concern in the following discussion is to show the special ways in which atomic and molecular physics is now contributing to biophysics.

The action of ionizing radiation on living systems is the subject of much current research. E. C. Pollard gives three major reasons for the value of these studies. The first is the expectation that an understanding of the interaction between a physical agent and living cells will lead to fuller knowledge of the structure and function of cells. The second is the effect that ionizing radiations have on cancerous cells. Under certain circumstances tumors can be cured or induced; to learn more about the factors that stimulate or retard their growth is essential. The third reason is that radiation is everywhere. We generate it; it is present in the cosmic radiation; and it arises spontaneously from the earth. Everyone is exposed to radiation. Therefore, it is vital that we develop a fuller understanding of the effects of radiation, man-made or natural, on life.

Some ionizing radiations are produced by radioactive decay of certain atomic nuclei and also by certain kinds of rearrangement of the electrons surrounding nuclei. In the former process, energetic particles are produced that can penetrate matter, leaving behind a trail of ionization. Accompanying some radioactive decay processes are gamma rays, which lose their energy either by absorption in a single nucleus or by interacting with several atomic electrons. This latter process involves a rearrangement of extranuclear electrons and is familiar as a major source of x rays and ultraviolet light. X rays and ultraviolet light also deposit their energy either actively in a single locale or in a few spots rather than by leaving a trail. At these spots, as is true also in the case of gamma rays, fast electrons are generated, and these leave trails of ionization. Some gamma rays generate electron–positron pairs that ionize the matter through which they pass. The positions may follow a complicated path leading to their ultimate disappearance and accompanied by ionizing radiations. Therefore, whatever the source of the primary radiation, the ultimate biological problem is to discover the effects of fast-charged particles on the atoms and molecules of living cells.

Because the effects of radiation on living tissues can manifest themselves by altering the functioning of organisms for the worse, the discipline of health physics has arisen. It can be regarded as either an independent discipline or a part of biophysics, depending on one's point of view. Health physics is committed in part to the protection of man and his environment against the harmful effects of ionizing radiation. In practice, the health physicist must first determine the properties of the radiation to which man or the environment is exposed. These determinations are made with instruments with which the radiation interacts. On the basis of these measurements a prediction of the biological effects that would be produced on a living system exposed to such radiation is possible. To extrapolate from physical measurements to biological effects requires a great deal of scientific input; nuclear, atomic, molecular, solid-state, and plasma physics, as well as radiation studies at chemical and biological levels, play crucial parts in this effort.

Atomic physics is closely related to the effect of radiation on dense matter, as the following example shows. The manner in which ions lose energy as they traverse a sample of matter is governed by a fundamental atomic property, the oscillator strengths of transitions in isolated atoms. The energy-loss processes result in excitation and ionization of the target atoms. The quantitative measure of energy loss in matter is called the "stopping power" of the material. This quantity determines the range of a particle of given energy in the material. Energy-loss distributions and ranges are essential quantities in radiation detection and dosimetry, in understanding biological effects, and in the treatment of cancer with ionizing radiation.

Beyond these relatively elementary applications lies the essentially unexplored domain of macromolecular structure and interactions as derived from first principles. This subject invites the attention and efforts of atomic and molecular physicists who are seeking fruitful areas in which to use the techniques and knowledge that are developing currently.

D. EDUCATION

Atomic and Molecular Physics in Undergraduate and Graduate Education There is little need to discuss in depth the role of atomic and molecular physics in undergraduate education. It is a pervasive theme in most undergraduate courses, from the general survey courses for nonscience majors to the advanced electives. A large percentage of experiments in undergraduate (and graduate) laboratories deal with atomic and molecular physics. The undergraduate physics major usually takes a laboratory course in his junior or senior year, in which he is exposed to experiments in electronics, experiments involving electron beams, determinations of the charge and the charge-to-mass ratio of electrons and ions, optical and mass spectroscopy, and the like. Currently, more sophisticated experiments, involving lasers, molecular beam apparatuses, and microwave and radio-frequency resonances are performed.

In addition to formal laboratory course work, atomic and molecular physics offers an interesting and often unique opportunity to the undergraduate student and professor to perform worthwhile and creative research, albeit on a small scale. Many of the best young people who enter the education profession want to remain active in research even though their primary responsibility is teaching. This motive causes them to maintain contact with physics throughout their teaching careers. The brightest undergraduates also are eager and willing to advance beyond the prescribed curricula to undertake some creative activity. For both groups, atomic and molecular physics is a fertile field of endeavor.

The kinds of research that a physicist can do at a small, and often isolated, college are limited. The equipment needed cannot be too elaborate or expensive, nor are experiments requiring the efforts of more than one or two people practical in this context. The kind of studies conducted under these limitations must be worthwhile, that is to say, both stimulating and publishable. The college teacher-researcher must choose problems that are significant but that do not compete directly with those studies at major centers. Finally, he should try to select a problem that is sufficiently significant to attract some sort of financial support. Though the college teacher-researcher will not need large sums of money, he should have an amount sufficient to buy modern equipment, provide a summer stipend, and hire one or more senior physics students to help him. In some cases, obtaining a few thousand dollars for summer employment makes it possible for the college teacher to afford the luxury of teaching.

But the research program the college teacher chooses must not only meet the requirements just described. The project must also be profitable as a teaching tool for promising science majors who should taste some of the satisfactions of doing research. The project must be such that a student who works for one or two summers and an academic year will profit from the research experience and, at the same time, contribute meaningfully to the advancement or completion of some phases of the work. Thus, the student must be able to comprehend fully the problem being investigated. The experimental equipment should not be so complex and intricate that the student spends most of his time learning about the apparatus and how to operate it. However, the research should employ equipment that is modern and similar to that with which the student will work as a professional scientist.

A look at graduate school course listings in physics and at the contents of typical qualifying examinations for advanced degrees shows that atomic and molecular physics plays as central a role in graduate education as in undergraduate. Therefore, the points made in regard to undergraduate programs can be generalized to graduate training.

Some questions arise concerning the current status of atomic and molecular physics as an active research area. One of the purposes of the atomic and molecular physics survey was to determine the status of research in this field. Perhaps the most significant feature revealed by the survey was that over 50 percent of the responding institutions have added new faculty in atomic and molecular physics in the last two years. The number of graduate students entering the field has shown an equal increase during the same period. The research areas showing the greatest increase in activity include: (a) atomic and electronic collisions, particularly using crossed-beam techniques, (b) optical spectroscopy, particularly relating to laser sources and beam-foil techniques, and (c) those areas most closely connected with chemistry, including reaction-product and reaction-rate studies, using crossed-beam, afterglow techniques; optical spectroscopy, including level-crossing, laser-induced transitions; and other resonance-fluorescence methods.

As survey respondents indicated, one of the most generally experienced problems in the typical experimental research group in atomic and molecular physics in recent years is the need to employ increasingly complex and sophisticated equipment and the corresponding increase in the cost of doing innovative research. Even in this "small science," it becomes increasingly necessary to employ elaborate data-processing procedures, fast-pumping and ultrahigh-vacuum techniques, expensive laser equipment, monoenergetic electron optics, and the like. Automation, still not generally fully accomplished in the typical atomic and molecular physics experiment, will increase. Thus the researcher in this field is faced continually with the need to obtain equipment funds that are too great to be extracted from typical capital equipment budget items in research grants.

The theoretical student and faculty member might be experiencing a similar problem. Although the employment of simplified models and suitable approximation schemes remains the central theme of theoretical atomic and molecular physics, the use of systematic expansion methods employing high-speed computers is growing rapidly. Thus, the availability of computer time on large computers represents a serious problem to the small research group, especially one not associated with a first-rate computational facility.

With modern equipment and computational facilities, the graduate student in atomic and molecular physics generally experiences an unusually wide-ranging education. Typically, he is involved in all aspects of his research problem, from the purely hardware aspects through data analysis and theoretical calculation. Throughout his research he encounters problems in a large number of specialties, giving him flexibility and a relatively broad base of training, highly desirable features of any contemporary educational program in the current rapidly changing employment environment.

10 Lasers and Quantum Electronics

A. ATOMIC AND MOLECULAR PHYSICS IN LASER DEVELOPMENT

The laser is a device for obtaining the coherent release of radiation in a pulse or in a continuous wave from large numbers of suitably prepared atoms or molecules. The laser output is characterized by its spectral sharpness, its high degree of directionality, and the very-high-energy densities that can be achieved in a focused laser beam.

The first lasers were crystals; they were excited by a flashlamp and emitted coherent radiation in a single powerful burst at a wavelength characteristic of some atomic or molecular constituent of the crystal. More recently, the gas laser, a gas-discharge device in which excitation depends on collision processes, has led to a continuous mode of operation and added greatly to the versatility of the laser as a technological device.

All lasers depend on some mechanism for pumping atoms into the upper level of the laser transition and producing an "inverted" population of energy levels rather than the thermal population, in which the populations would conform to the Boltzmann relation with the larger densities in lower levels. Collisional processes must be such that the upper state is preferentially populated. In the continuous-wave (cw) laser, the lower state must be efficiently depopulated. A reasonably complete knowledge of the rates of populating and depopulating processes and the competing radiative transition rates obviously facilitates the discovery and development of lasers for various purposes.

Unfortunately, the "engineering tables" of these atomic collision properties are not available. Tables on the structural properties of atoms (NBS Circ. 467, *Atomic Energy Levels*) are fairly complete and extremely valuable for this purpose. Tables of radiative transition probabilities are being developed around existing data, and the research effort in this field has increased substantially as a direct result of the requirements of laser development. There are some systematic efforts to measure collision properties in direct support of laser development. However, the general approach to laser development is not to design them but, by highly empirical investigations, to discover systems that support laser action. Under present conditions, the attempt to understand such systems in order to optimize performance is severely limited by the lack of data.

Two important laser systems that have been extensively studied and analyzed will be discussed briefly to illustrate the relationship of atomic and molecular physics data to the gas laser.

The pulsed argon ion laser, operating in the 4756-Å (blue) transition, is represented by the energy-level diagram in Figure 6. Operation in the pulsed mode seems to depend on a direct excitation by electron impact to the $4p\ 2P^{\circ}_{3/2}$ state of Ar^{+}. The excitation cross section for the lower state of the laser transition must be smaller than that for the upper state to understand the population inversion that is achieved. This difference in the excitation cross sections has been qualitatively confirmed through "sudden"-approximation calculations.

Continuous-wave operation of the argon ion laser is also possible and has been the subject of extensive studies. The excitation apparently consists of a multistep process, ionization with subsequent excitation, but the details are not well understood. Short radiative lifetimes of the lower state keep that state depopulated.

The CO_2 gas laser, now attaining powers in excess of 1 kW, with 15 percent efficiency in continuous operation, is understood in terms of a quite different and more complex set of excitation and deexcitation mechanisms. In one form, the discharge medium is a mixture of CO_2 and N_2, and it has been established that the excitation of the upper level of the laser transition comes about through resonant energy transfer from vibrationally excited N_2 in the ground electronic state. Only relatively recently has the necessary high concentration of N_2 in excited vibrational states been understood. Experimental

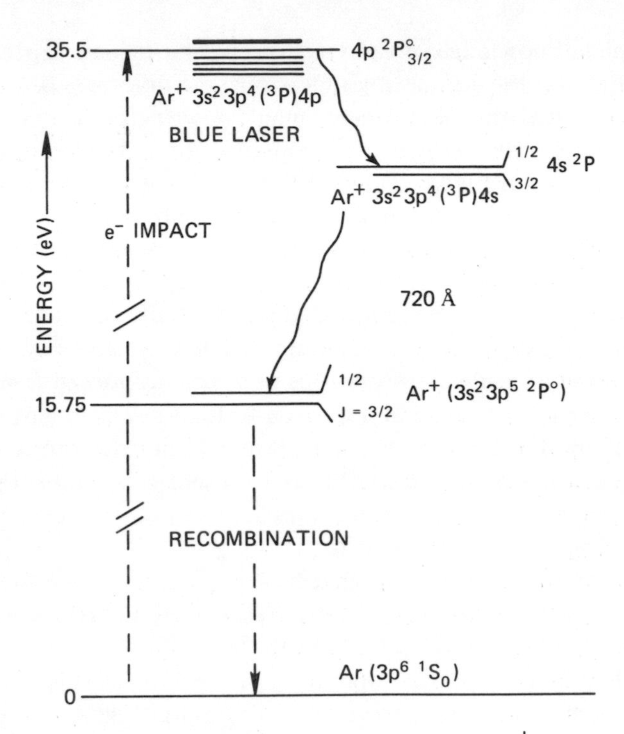

FIGURE 6 Energy levels pertinent to the Ar^+ laser excitation process. Only the high-gain 4765-Å laser transition is specifically indicated.

work using high electron energy resolution has shown that the cross sections for direct electron impact excitation of the vibrational levels are unexpectedly large and sharply resonant. The interpretation of these results is in terms of the formation of short-lived negative ion states of N_2^- via the temporary association of the incident electron with vibrationally excited N_2.

The CO_2 laser is also operable in the absence of N_2. In this case, the excitation role is played by CO produced by dissociation under electron impact. Large vibrational excitation cross sections for CO are experimentally observed; they are interpreted in terms of temporary states of CO^-.

Efficient destruction of the lower levels of the laser transition is required to maintain population inversion and avoid limiting the available laser power. In the CO_2 laser the destruction process is collisional, the lower levels, consisting of vibrational bending modes, being much less stable in collision with N_2, CO, or He than are the upper levels. The addition of helium not only aids in the destruction of the lower state but is also effective in lowering the gas temperature, thereby substantially reducing collisional loss of the upper state while only slightly reducing destruction of the lower state.

The CO_2 laser and the argon ion laser provide illustrations of the many ways in which microscopic atomic interactions can play vital roles in laser technology. Other types of laser depend on other processes. The gas laser is, in fact, simply a sophisticated form of gas discharge, which differs from more conventional discharges in the introduction of an optical resonator, the necessity for appropriately large interaction volumes, and atomic interactions that can operate on a time scale of a few nanoseconds. It presents some new and challenging problems to atomic physicists, some of which are diagnostic problems. Other problems relate to the need for design data, such as critical cross sections. Ultimately, supporting research in atomic physics must provide the data necessary for the design of new types of laser as well as the refinement of existing types to permit specific technological applications.

B. QUANTUM ELECTRONICS

Gas lasers belong to the even broader area of quantum electronics, defined as that area of electronic technology, particularly spectroscopy, in which quantum-mechanical processes are directly responsible for the relevant technology application. Quantum electronics is providing radically new techniques and devices in such diverse fields as chemistry, optics, and astronomy. Research activity is great in these interrelated fields; therefore, they attract some of the most talented scientists. Those Nobel laureates who have received their awards for work in spectroscopy and quantum electronics in recent years provide examples: P. Kusch, W. E. Lamb, F. Bloch, E. M. Purcell, N. G. Basov, A. M. Prokhorov, C. H. Townes, and A. Kastler.

Although quantum electronics is recognized as a separate discipline, it is inherently interdisciplinary, drawing on and contributing to solid-state, atomic, molecular, and chemical physics, as well as electrical engineering. One example of an entirely new phenomenon properly belonging to quantum electronics is nonlinear optics, which was hardly conceivable without the availability of intense laser light. When nonlinear optics was first studied, adequate understanding through reasonable extensions of classical optical theory was anticipated. However, there were many surprises, including the discovery of self-focusing, that is, the tendency of a high-intensity beam of light to focus itself in passing through a medium. Among the most interesting recent discoveries have been the methods for producing picosecond pulses of high-intensity light, with the confident expectation that still shorter and stronger pulses will soon be obtainable.

Very short light bursts produce a range of new optical phenomena. These include induced transparency (roughly speaking, this results from the inability of the atomic electrons, because of their inertia, to respond to optical stimulation) and photon echoes, which are in a sense true echoes that result from the delayed response of a radiating system to an optical stimulus.

C. THE USE OF LASER METHODS TO STUDY ATOMIC PHYSICS PROPERTIES AND SYSTEMS

Lasers may offer the largest dynamic range usually encountered by physicists; there are laser intensities beyond 10^{12} W/cm^2 and receiver noise levels easily below 10^{-12} W. The time–spectral width continuum is possibly more interesting to atomic physicists, since many of the fundamental discoveries in their field have been made with spectroscopic methods. Thus with pulse durations of a few times 10^{-12} sec and cw sources with frequency instabilities well below one part in 10^{12}, one can anticipate profitable application to the atomic physics problem of both impulse (time domain) experimental methods and cw spectroscopic (frequency domain) methods. Let us illustrate this situation by considering a specific case, that of the internal-state physics of liquids.

Atomic physics began as an analytical science. Progress resulted from discussing problems on an individual particle level, allowing only two-body forces and treating systems with a small number of particles, generally no more than two. With the insight gained from this work, several more or less natural approximations could be made, which permitted significant investigations of more complex systems (heavier atoms and small molecules) using numerical methods. However, it was obvious that extrapolation of these methods to the density levels of solids or liquids was not attractive: the experimentalist could specify the relevant condition of his system by at most a few tens of parameters—not 10^{21} or 10^{22}. Similarly, the fantastic detail produced by calculations had to be suitably integrated and averaged, thus losing the large bulk of the individual particle effects, although retaining the effects shared by many particles. The introduction of the concept of "elementary excitations" of the system allowed a quantum step forward in the description of these many-body effects. These modes, of course, are merely linear combinations of the single particle states, but, because of their lower energies and useful symmetry properties, they proved to be "natural" coordinates for the problem. Great progress in superconductivity theory was brought about by the invention of the "Cooper pair," the spin-paired two-electron fundamental building block of the theory. In magnetic solids, the concept of "magnons," localized zones of reversed magnetic moment, serves as the simple excitation that enables succinct description of the magnetic behavior. At the moment, the frontier of this work is, perhaps, in the physics of the liquid state. We do not yet know the simple way to think about liquids, but two laser experiments recently have taught us some essential new facts that will have fundamental effects on our intuition and understanding. Using mode-locked laser pulses of picosecond duration, the orientational relaxation of molecules in a liquid has been studied. The high intensity of the incident pulse creates an anisotropy in the liquid—molecules whose transition moments are parallel to the light vector are more strongly saturated by the gigawatt beam. Two weak-probe beams of orthogonal polarizations pass through the sample after an adjustable time delay. The ratio of the absorption of the two-probe beams thus gives information on the degree of orientational relaxation that has occurred during the time-delay interval. It was previously supposed that the orientational motion of molecules in a liquid would be strongly influenced by the formation of intermolecular "bonds," the temporary sharing of a hydrogen atom between two adjacent molecules. However, it is found that the viscosity-scaled orientational relaxation of the laser-affected solute molecule is essentially independent of the choice of solvent, a strong hydrogen-bonding agent (methanol) giving the same orientational relaxation rate as a nonbonding solvent (chloroform). Thus it appears that the reorientation process is dominated by the fluctuational occurrence probability for production of a local vacancy, rather than by the details of the intramolecular bonding. The experiments also show that the intermolecular bonds must form and break on a very short time scale.

Laser methods also have taught us recently some rather surprising things about shearing processes in liquids, using, in this second case, frequency-domain analysis with cw sources. It was commonly found that usual liquids do not sustain shearing forces on a time scale long enough for our physiological receptors to observe it. The dogma that liquids do not sustain shear stresses resulted. Brillouin scattering experiments in various liquids recently have been performed. The frequency shift of light scattered at a definite direction from the existing radiation is determined; thus, information can be obtained on the "velocity of the phonon mode" responsible for the light scattering. These methods give both the shift and the line width, yielding both "phonon velocity" and damping length. Therefore, one finds that liquids do, in fact, have high-frequency shear waves. L. Brillouin came to this conclusion many years ago by comparison of the specific heats of liquids and solid phases. The shear modulus is typically one-hundredfold lower than the elastic or bulk modulus. Damping times of 10^{-11} sec are found for these shear excitations of frequency 10^9 Hz. It appears that one must think of these shear waves as propagating modes. The velocity is about 500 msec, so the damping length is only 50 to 100 Å. However, the wavelength of the disturbance is approximately 3000 Å. Thus we begin to perceive that liquids usually have some kind of temporary, moderate range order. Perhaps "liquid crystals" are not so odd after all. Possibly, for some reason, their shear properties merely have longer relaxation times, by a factor such as 10^{10} or more, than those of ordinary liquids.

These two applications of laser methods to the study of atomic physics might appear to be overdramatic, because what we are learning is qualitatively new. We are just learning how to think about these problems. In many other areas, we have longer experience using other methods and, to some extent, already know what questions to ask. Such investigations are no less worthwhile than those that break entirely new ground. Indeed, when a field is ripe for the introduction of a new tool, the progress can be spectacular. The remarkable progress in critical-point physics in the years following the development of laser light-scattering methods provides an example. Fundamental questions on the order of thermodynamic transitions were posed and answered. A new type of elementary excitation was discovered experimentally and a sound, unified theoretical explanation developed. Stimulated light scattering was observed from these thermal excitations (Rayleigh wing scattering).

Raman scattering is another classical field that was exploded into activity by the introduction of laser methods. The fantastic concentration of monochromatic power made possible by laser coherence gives in minutes Raman spectra of gases at atmospheric pressure. In the prelaser era, it was not uncommon to use exposures of days with solid and liquid samples at densities three orders of magnitude higher than the gas-phase studies now possible. In fact, the introduction of appropriate laser methods may end the use of Raman spectroscopy as a field of atomic physics. Commercial instruments have become available, and analytical laboratories are routinely using Raman as well as infrared spectroscopy to determine compositions and the like. Perhaps, for the atomic physicist, laser Raman spectroscopy will become primarily a highly useful tool, such as digital and analogue electronics or conventional infrared, visible, and ultraviolet spectroscopy.

Probably because of physicists' long experience in the use of spectroscopic methods to investigate atomic physics problems, these areas have been the first to expand with the introduction of the new laser techniques. It is interesting that large-scale digital data-acquisition methods and laser spectroscopy arrived in the laboratory at roughly the same time. Consider, for example, the spectroscopic study of molecular spectra. Because of the multiplicity of internal modes, the spectra of molecules rapidly become very rich. For molecules of three or more atoms, conventional spectroscopy begins to run into resolution problems. However, the more significant limitation is that the number of lines is too great for traditional line-classification methods to work. Dependence on human inspection and inspiration was necessary to detect the regularities that ultimately make possible the unraveling of the spectra and assignment of atomic (molecular) levels. One could imagine using computer search methods to establish equalities of energy intervals and thus to aid the spectroscopist in perceiving the regularities in the spectrum. In fact, such methods have been employed at the National Bureau of Standards in analyzing the spectra of rare earth ions. However, it remained for the monochromatic laser to advance the field of visible molecular spectroscopy. The usual gas laser sources are sufficiently monochromatic that they typically will pump zero or one rotation-vibration component of some visible absorption band of a gas-phase absorber. The fluorescence spectrum from the single excited state thereby prepared is, of course, much more simple than normal molecular spectra, which represent a superposition of enormous numbers of these single spectra. Thus, in these fluorescence spectra from a single excited state, one can see the regularities and structure associated with the ground state. Identification of the spectral transitions leads to derivation of the half dozen or so molecular constants that describe the single excited and multiple lower states. Other laser lines pump other excited states. With transitions of known assignment, the classical methods of atomic physics, such as the Hanle effect, level crossing, double resonance, and the like, can be employed to elucidate the structure further.

This kind of work is pursued in several laboratories on many different molecules: The "accidental" laser line–molecular absorption line coincidence is not especially rare. However, the use of tunable organic dye lasers in all probability will be the critical element that brings this kind of spectroscopy to a high level of activity.

The dye laser offers a new ease and versatility in the investigation of spectral resolution. By a reduction in intensity of roughly a factor of 3, it is possible to narrow the radiation bandwidths of the dye laser from hundreds of angstroms to the milliangstrom level implied by the duration of the laser's excitation pulse. The dye laser pulse duration similarly can be varied from the 5-nsec region, using the pulsed nitrogen laser for excitation, to the microsecond region, using flashlamp excitation, to complete cw operation, using a cw laser such as the argon ion laser as the exciter. This unrivaled versatility will result in remarkable spectroscopic advances. Already, survey studies are showing interesting changes in molecular fluorescence lifetimes (in NO_2) as a function of the vibrational quantum number. Diffraction-grating-narrowed dye laser radiation has been used to scan the energetic threshold for the detachment of electrons from the negative ion of sulfur, providing the first electron affinity value of submillivolt precision. Dye lasers have been used also to probe atomic energy levels of potassium that were distorted from the free atom values by intense ruby radiation and to detect a natural layer of atomic sodium at an altitude of 95 km. The dye laser clearly offers the promise of high-resolution spectroscopy in situations that will be of great significance in making air-quality measurements. Resonance phenomena will allow specific impurities to be detected remotely and monitored. Possibly, the dye laser constitutes a combination of utility, simplicity, and economy that will cause laser methods to have an impact on nearly all areas of physics that are related to optical phenomena.

D. SATURATED ABSORPTION AND ULTRAHIGH-RESOLUTION SPECTROSCOPY

Because of the high intensity and narrow bandwidth of suitable laser sources, it is possible to achieve spectroscopic resolution vastly in excess of the Doppler-effect-limited resolution previously available. (It is important to realize that this resolution limit was not an instrumental limit; rather, it arose from the thermal excitation of the translational degree of freedom in the gas phase.) With two laser fields suitably applied to the gaseous sample, a nonlinear resonance feature can be observed, the width of which can be orders of magnitude less than the Doppler line widths. For example, saturated absorption resonance line widths of less than 45-kHz full width have been observed on a methane transition at 3.39-μm wavelength. This resolution of 2×10^9 is probably the highest value yet reported in optical spectroscopy and rivals the resolution available with hydrogen beam storage techniques in the radio-frequency domain. The application of these saturation methods to spectroscopy is being pursued vigorously by a number of laboratories. The national standards laboratories are studying the possible definition of a wavelength standard based on this phenomenon. An interesting sidelight is that the rate of progress in this area is so great that the concept of a wavelength standard is likely to become obsolete, for several laboratories already have achieved direct heterodyne frequency measurements into the 10- and 5-μm wavelength region (30 and 60 THz). Probably, the frequency of a suitable stabilized laser will be measured against the cesium frequency standard. The wavelength of the methane-stabilized laser has been measured against the present internationally accepted standard of length, the krypton-86 line at 605.7 nm. The (inverse) ratio will be the velocity of light, which will then have been measured with an uncertainty dominated by the uncertainties in the krypton length standard. Many people believe that it would be desirable then to define the velocity of light and thus to eliminate the use of an arbitrary primary length standard. Obviously, in practical length measurements, it might be convenient to continue to use a set of secondary standards, probably (saturated absorption) stabilized lasers the wavelengths of which will be established by precision interferometry. Thus the situation in length metrology will be roughly comparable to that in the definition of the time scale, in which uniform atomic time is useful in precision laboratory work, but the use of a (less stable) earth-based time scale is more valuable in astronomy and some kinds of space research.

It is impossible to be all inclusive in this brief chapter. Many exciting and promising applications of lasers to atomic physics research must be omitted, for example, the use of injection diode lasers and Raman spin-flip lasers to study infrared molecular spectra. However, it is likely that some of the most interesting applications will be in qualitatively new areas—areas that are only imperfectly perceived at the present time. Possibly, the next significant advance will be the development of a coherent x-ray source, or the application of laser methods to the "classical" relativity experiments, such as the Michelson-Morley experiment, might reveal a new realm of added richness in physics.

11 Atomic and Molecular Physics in Technology

The microscopic behavior of free atoms, molecules, ions, and electrons determines their behavior in bulk matter. Therefore, understanding of these microscopic processes is one starting point for a rational solution to technological problems involving the gas phase. Some uses of such atomic and molecular physics-based technology are found in the communications industry, environmental studies, weather forecasting, the fluorescent and electric arc illumination industry, gas lasers, vacuum and ultrahigh-vacuum technology (an entire industry stimulated and abetted by atomic physics basic research), aerospace problems, including re-entry and communications, rocket propulsion, rocket wakes and fuel technology, military physics, including problems of defense against nuclear attack and the effects of nuclear bursts in the atmosphere, the search for controlled thermonuclear fusion, and plasma and electric discharge devices, such as magnetohydrodynamic generators and high-intensity flashlights. From the many examples of the relevance of atomic and molecular physics to technology, we have chosen a few to describe in detail.

A. FUSION RESEARCH

For two decades there has been a worldwide effort to harness the fusion process for the purpose of obtaining a virtually limitless and relatively noncontaminating source. This effort is centered principally in the United States, the Soviet Republics, Great Britain, Germany, and France. The demand for electric power could soon reach crisis proportions; therefore, research on controlled thermonuclear reactors has received increasing emphasis.

Many difficulties have beset the fusion quest; promising approaches were frustrated by unforeseen problems, and periods of optimism often were followed by disappointment. The success of the Soviet Tokamak machine has led to renewed expectations that adequate containment times and temperatures will be attained. Several such devices are under construction at AEC laboratories in the United States.

The close relationship between the fusion program and atomic physics can scarcely be overemphasized. Atomic and molecular physics plays a role in formation of the plasma, loss mechanisms that cause instabilities, and plasma diagnostics.

From a historical standpoint, the research in the creation of fusion-type plasmas has centered around two different schools of thought. One claims that the problems are less difficult if a cold gas is heated by various means to a thermonuclear temperature (10–100 keV). The other argues that it is necessary to inject high-energy (10–600 keV) ions or atoms into some magnetic configuration to create a high-temperature plasma. There is agreement that the plasma must be contained in a magnetic environment, and two concepts have evolved for the containment geometry. One method places the plasma in a closed-type magnetic field, such as a toroidal field, where scattering losses are minimized. The other method is an open-end-type geometry, using magnetic mirror fields; by optimizing various parameters, a high-density, high-temperature plasma should be obtained.

Atomic scale properties, the yardstick of which is the angstrom (10^{-8} cm), must be contrasted with the nuclear scale process of a controlled fusion reactor, for which the yardstick is the fermi (10^{-13} cm). However, since the probability of particle reactions involves approximately the square of these quantities, the atomic process rates must be reduced by possibly ten orders of magnitude for the nuclear processes to dominate! According to this concept, the atomic and molecular loss processes can be used to define a figure of merit, Q, for a fusion-type plasma. Q is defined as the ratio of nuclear power produced to the particle power lost by atomic processes.

The nuclear power produced is the product of the square of the charged particle density (n_+^2), the nuclear reaction rate parameter (σv) (where σ is the cross section for the desired nuclear reaction and

v is the mean particle speed), and the energy produced per reaction (17.6-MeV reaction). Similar calculations define the power lost by atomic processes as the product of the charged-particle density, the neutral-particle density in the plasma region, the atomic reaction rate parameter $(\sigma v)_A$, and the average energy of the plasma ions. The performance of thermonuclear reactors reveals a limitation. If, for example, only one atomic loss process (such as charge exchange) is involved, then the density of the neutral deuterons or tritons in the plasma must be seven orders of magnitude less than the charged-particle density. If by chance the plasma cannot be characterized by a true temperature, that requirement can increase to ten orders of magnitude. Clearly, one has to have a good idea of the actual energy-loss processes to design an efficient reactor.

Several examples illustrate the dominant role that atomic processes have played in the development of our ability to produce and isolate high-temperature, high-density plasmas. The formation of plasma from an initially cold neutral gas and its confinement in a toroidal magnetic field depend critically on atomic cross sections. The limitation on plasma heating, and its thermal isolation, in the United Kingdom "Zeta" experiment (*circa* 1957–1960), was the absorption of most of the input energy by ionization and excitation of impurity oxygen atoms, followed by its radiation as resonance O^{+5} radiation. Similar results were found in the Princeton Stellerators, although the realization of the importance of impurity atoms made the reduction of impurities an important part of the design of these devices. The impurity level in the Model C Stellerator (1961–1969) was finally reduced to ~0.1 percent by the installation of magnetic divertors to carry the magnetic flux at the surface of the plasma and the attached impurity ions into a separate vacuum system. Only the influx of hydrogen atoms from the vacuum system walls remained to limit the plasma temperature.

The Model C Stellerator could heat plasmas of density $\sim 10^{-13}$ cm^{-3} so that $kT_3 \sim kT_i \sim 100$ eV by heating, first the electrons and then, by collisions, the ions. If the ions were then separately heated by ion cyclotron resonance heating, a value of $kT_i \sim 500$ eV could be achieved. In both cases, the confinement time of the plasma was ~1 msec. The U.S.S.R. Tokamak T-3A has reached electron temperatures in excess of 1000 eV, ion temperatures of approximately 500 eV, and confinement times of the order 10 msec; similar (and somewhat higher) figures have recently been achieved with the Princeton Model ST Tokamak (which is Model C rebuilt). Both the Model C and Tokamak results and limitations seem explicable in terms of the atomic processes and the known impurity concentrations; therefore, it should be possible, in principle, to construct a larger device that will reach the fusion regime. In all these studies, a detailed knowledge of the atomic processes involved was essential for plasma formation, heating, and diagnosis.

The importance of atomic processes has been clearly demonstrated in the attempts to obtain a fusion plasma by injecting energetic particles (10 to 600 keV) of deuterium atoms or diatomic deuterium ions. Initially D atoms must be formed in large quantities, either by charge exchange of D^+ in gas cells or thin films or from the dissociation of D_2^+. In addition, theory and experiment indicate that electrostatic instabilities in the plasma result from anisotropic velocity distributions in the injected particles. Thus the additional requirement has been imposed that the injected beam have a controllable velocity distribution. As energetic particles enter the magnetic field, they must be trapped in the magnetic bottle by some irreversible processes such as ionization of D° by residual gas, plasma dissociation of D_2^+, or the Lorentz ionization of D°. All trapping processes are inefficient, and a great need exists for improved methods of trapping. These injection techniques have been extensively exploited by several countries (Russia, Ogra; United Kingdom, Phoenix; United States, DCX's and Alice), but so far the density has been limited to less than 10^{10} particles/cm^3. This limit results principally from the inability to inject and trap large currents of deuterium, the difficulty of overcoming the charge-exchange loss of contained deuterons, and the presence of general plasma instabilities.

The present attempt to inject large fluxes of particles into a plasma through the use of macroscopic clusters of deuterium molecules provides an example of the continuing need for understanding basic atomic and molecular parameters. Preliminary estimates indicate that several amperes of equivalent neutral particles can be formed as clusters through expansion of deuterium gas at low temperatures. Little information is available about the cluster formation, the intermolecular forces that hold the cluster together, and the ionization and dissociation of the cluster. This problem is only one of many that face the atomic physicist in fusion research.

The many atomic and molecular properties that affect the creation of a high-temperature, high-density plasma merit serious consideration. Such processes as ionization and dissociation trapping (including the esoteric Lorentz trapping mechanism and inverted cascading), recombination and charge transfer, neutralization in gas and vapor cells, multiple scattering, and stopping-power or energy-transfer losses need further study. The properties of surfaces in relation to absorption, outgassing of cryopumping surfaces under particle bombardment, sputtering, and beam disposal, as well as atomic and molecular state lifetimes and oscillator strengths, have to be better understood.

As always it is difficult to project the future requirements and importance of atomic processes as the plasma physicists advance toward the ultimate goal of a fusion reactor. When the fusion reactor becomes a reality, one will have to solve such problems as feeding fuel particles into the inner volume of the reacting system, removing impurities from container walls, and maintaining neutral particles inside the plasma volume.

B. ATOMIC AND MOLECULAR PHYSICS IN SPACE AND DEFENSE TECHNOLOGY

Many problems of national defense as well as those of the spaceflight program involve advanced technology based on atomic and molecular physics. A large class of phenomena and processes deals with (gaseous) systems with a high-energy density, which corresponds to a high temperature. The temperature range of 1000–10,000 K, or 0.1–1 eV, encompasses a number of these problems. Additional phenomena of technological significance occur in much higher-energy ranges corresponding to temperatures of several million degrees—which is comparable to the range of phenomena encountered in stellar atmospheres and plasmas.

As the temperature of a molecular gas increases above approximately 1000 K, drastic changes take place in the internal state of the gas. We no longer have an ideal gas characterized by a ratio $\gamma = c_p/c_v = 5/3$ for a monatomic gas and 7/5 for a diatomic gas (where c_p, c_v are the specific heats of constant pressure and volume, respectively) but gases that display specific and characteristic excitation of rotational, vibrational, and electronic degrees of freedom, both bound and free, such as correspond to dissociation and ionization. There is a significant difference here between bound motions and free motions. For discrete energy levels, the population n_{exc} of an excited state in local thermodynamic equilibrium, relative to the population n_0 of the ground state, is characterized by the Boltzmann relation

$$(n_{\text{exc}}/n_0) = (g_{\text{exc}}/g_0)\, e^{-E_{\text{exc}}/kT},$$

where E_{exc} is the energy of the excited state relative to the ground state, kT is the thermal energy, and g is the statistical weight of the corresponding level. There is significant excitation of the upper level for $T \geqslant E_{\text{exc}}/kT$. On the other hand, the relative population of ionized states is characterized by the Saha equation (or its analog for a dissociation),

$$n_{\text{el}} n_{\text{ion}}/n_0 = G(m, T, g)\, e^{-I/kT},$$

where n_{el}, n_{ion} are the electron and ion densities, I is the ionization energy of the neutral particle, and G is the partition function. This leads to the result that for pressures between 10^{-3} and 10^3 atm there is significant dissociation, excitation, and ionization under suitable combinations of density and temperature.

Representative energies for normal diatomic molecules are

E(rotational)	10^{-2} eV ~	100 K,
E(vibrational)	0.2 eV ~	2000 K,
E(electronic)	5 eV ~	50,000 K,
E(dissociation)	5 eV ~	50,000 K,
E(ionization)	10 eV ~	100,000 K.

Because of the relatively high probability associated with the ionized state, that is to say because $G >> g$, typical rotational, vibrational, and excitational temperatures are as listed above, but typical dissociation and ionization temperatures are a factor of 10–15 lower than those listed above. Thus one observes that over the temperature range 1000 K to $T < 10{,}000$ K there is a great deal of action, involving the excitation of many degrees of freedom, a great deal of chemical rearrangement and bond-breaking, a large energy absorption or effective heat capacity, many highly specific optically and electrically observable phenomena, and a vast range of plasma phenomena.

At temperatures below 1000 K most gases are essentially "ideal" at normal pressures. Above this temperature lies the domain of greatest interest and challenge. One cannot overstress the need to develop our high-temperature technology to open this area fully to scientific study. The following sections discuss some areas already under intensive investigation.

C. MISSILE AND SPACE VEHICLE RE-ENTRY

A typical re-entry speed is 7 km/sec; the peak heating on a vehicle occurs at the "stagnation point," a blunt portion at the front of the vehicle where all the kinetic energy corresponding to the relative motion of 7 km/sec is transformed into heat. The "stagnation temperature" is about 7000 K in air, as compared with 16,000 K in argon (which is ionized but not dissociated) or 30,000 K for a $\gamma = 5/3$ monatomic ideal gas. The differences among these three temperatures indicate clearly the importance of dissociation and ionization as opposed to kinetic energy of translation in providing a reservoir for energy.

The stagnation region corresponds to the highest temperature, but (because of the long residence time of the gas) it also corresponds generally to local thermodynamic equilibrium (LTE) conditions. The boundary layers along the sides of the vehicle, corresponding to the region of frictional wall heating, or in the wake, which corresponds to the dissipation of the inhomogeneity induced in the gas that has been in thermal or mechanical contact or both with the vehicle by interaction with the undisturbed atmosphere, are in general lower temperature environments ($T \sim$ 2000–4000 K), but in both there are large nonequilibrium effects.

The problems of predicting and simulating full-scale re-entries require the use of partial simulation techniques, experimental techniques that range from cross-beam measurements of collision cross sections to shock-tube studies of gases at high temperature and from hypersonic gun ranges to specially designed and instrumented full-scale flight tests.

D. COMBUSTION

A typical rocket combustion chamber temperature is 2000–3000 K, so that as a result of chemical rearrangement there is a significant amount of vibrational excitation. There are also many free radicals and other unstable species. In the rocket exhaust, we consider the expansion of the combustion products through a nozzle—leading first to a cooling and freezing of the chemical forms present and subsequently to an interaction of the products with the ambient atmosphere—at a relative speed of 2–3 km/sec, so that further excitation and interaction occur. Studies of this interaction have major implications for jet aircraft design.

E. NUCLEAR FIREBALL

A nuclear fireball results when the air surrounding a nuclear explosion is heated by the x rays of a bomb. The air then expands and rapidly loses energy by radiation. The effective temperature of the hot fireball drops quickly to 6000–8000 K, but at this point the effective emissivity becomes so small that the cooling rate slows down. The environment of a highly nonequilibrium gas at temperatures of 6000 K and below, and containing a significant degree of ionization, lasts for an extended period.

F. HIGH-TEMPERATURE TECHNOLOGY

All the previous examples correspond to the same general pattern of nonequilibrium gas in the temperature range $1000\ K \leqslant T \leqslant 10{,}000\ K$. To have sufficient understanding of these phenomena for engineering or systems applications, it is necessary to have some background knowledge of the atomic and molecular physics problems of the particular environment. In this relatively low-energy range, present understanding is less than in the keV–MeV energy range, which has been studied much more extensively. It should not be inferred that high-energy (keV–MeV) atomic physics is not important for defense and space problems; rather, over-all understanding is greater, thus it is frequently possible to apply existing knowledge. This situation is not true of the low-energy range in which more work is needed to understand problems such as those encountered in re-entry, rocket combustion, and the atmosphere following a nuclear explosion.

A common feature of these situations, all of which involve heating of the atmosphere to a temperature in excess of 1000 K, is the reaction of an ionized gas to electromagnetic radiation. There are vast implications for the technology of communications; for example, in communication with a manned spacecraft during re-entry, detection of rocket wakes, and communication through the region of the atmosphere heated by a nuclear explosion. Currently, atomic and molecular data for the systematic analysis of these problems are not available.

G. AIR POLLUTION RESEARCH

Atomic and molecular physics is essential in achieving an understanding of the atmosphere. A vast array of interactions, involving photons, electrons, atoms, and simple molecules, determines the observed response of the atmosphere to solar radiation, cosmic rays, and man-made energy releases of various sorts.

With the large-scale introduction of organic and inorganic chemicals, for example in agriculture and military applications, the interactions become much more complex, both from the point of view of the numbers of competing processes and the molecular structures that have an important role. Historically, the study of air pollution has been more nearly in the realm of the chemist than in that of the atomic and molecular physicist. The critical problems have been associated with the chemistry of complex reactions, emphasizing both photochemistry and biochemistry.

But the chemistry of the pollution problem will not be fully understood until the chemical dynamics (atomic and molecular physics) of the atmosphere are understood. Such understanding requires the investigation of photoionization processes, photodissociation processes, and a great variety of collision reactions. Recent progress has been spectacular, but much remains to be done.

Since the chemistry of air pollution represents an extrapolation of the chemistry of the unpolluted atmosphere into more complex systems, it is not surprising that the techniques of pollution observation and study are often those first developed and applied in atomic and molecular physics. Recently, optical methods of analysis have become increasingly important. A wide range of the electromagnetic spectrum is of interest, but present activity focuses on discrete wavelength regions between 0.2 and 1.0 μ. In particular, ultraviolet and visible spectrometry and infrared and Raman spectroscopic methods are being investigated. These can be used in active or passive systems in which the transmission, backscattering, or emission of energy occur.

Laser radar systems are used to estimate atmospheric particulate concentrations and to study contaminant plumes. A variety of laser systems are under consideration for use in long-path optical monitoring systems. The development of tunable lasers and better coverage of the infrared spectrum will greatly enhance the prospects for obtaining definitive analysis from laser systems.

Other types of analysis, notably mass spectrometry, are also of great value. New techniques, based on the recent advances in the use of monoenergetic electron beams and the analysis of electron energy losses characteristic of the various pollutants, also are being tested.

The high specificity and sensitivity needed for analysis of air pollutants demand the development of analytical instrumentation. The expertise available in the community of atomic and molecular physicists will play a large part in meeting this need.

H. ELECTRIC DISCHARGES

One of the major efforts in atomic and molecular physics has been to understand the process of electric conduction in gases, which is the basis of a large number of technological applications described as "gaseous electronics" or gas discharges. The laser gas is perhaps the most sophisticated of these applications, but the production of light in less-advanced devices is still one of the most important technologies for our society. Light sources vary from milliwatt neon glow tubes to the ubiquitous fluorescent lamp, to kilowatt mercury and sodium arc lamps for street and public lighting, to the intense carbon and xenon arc lamps used in motion-picture projectors and searchlights.

Gas discharges are the basis for many electrical control devices. The regulator tube, so important in electronics, is an obvious example. The T-R (transmit-receive) switch, which is a key element in radar systems, is less well known but of comparable importance.

Atmospheric gas discharges, including lighting, have many technological applications. Atmospheric breakdown at high electrical frequencies limits the power that can be radiated from a radar antenna. High-voltage breakdown and loss of power by corona discharge are serious problems for ultrahigh-voltage transmission lines. Breakdown is a limiting factor in the generation of electric power obtained by separating ions by aerodynamic flow. It limits the voltage that can be used in x ray tubes, nuclear accelerators, and vacuum-tube rectifiers.

I. SURFACE AND EMISSION PROPERTIES

The entire vacuum-tube and electronics industry, prior to the time when the transistor was developed, was based on the early studies of thermionic emission. The pioneering work using clean metals in high-vacuum experiments provided the fundamental understanding of the emission process from which the empirical art of manufacturing coated cathodes with low workfunction later developed.

Experimental research in field emission gave the world the field electron and field ion microscope, used widely to study the structure of crystal surfaces. Field emission tips using much the same emission properties now supply high-power x-ray flashtubes.

The photocell and photomultiplier in its various forms developed from basic studies on the yield of secondary electrons resulting from primary electron impacts on various surfaces. Without these high-gain, low-noise amplifiers, innumerable industrial and scientific measurements and detection operations could not be performed.

Experimental research showed in 1922 that electrons have wave properties that are demonstrated in the way that electrons scatter from a crystal lattice. This early discovery had enormous implications in the development of physics, and recently it has provided a very tangible result in the widely applied low-energy electron diffraction (LEED) technique for studying the properties of surfaces. Widely used commercial apparatus using this principle is now available.

Experimental studies of the emission of electrons from a surface by photon absorption began even before Einstein's early work in this field. There are innumerable applications of photoemitting cathodes for industrial and scientific purposes. Recent work extending the sensitivity of such cathodes further into the infrared for night vision and infrared imaging has been pursued in both military and civilian laboratories. The television tube resulted directly from such work.

The various types of ion engine now being developed and tested, which will become increasingly important, originate directly from phenomena and techniques well known in atomic and ion beam studies in plasma work or in thermionic emission work. Ions are generated by electron impact, or by heating vapor in contact with reactive cathode surfaces, and accelerated by electrostatic fields. Magnetic force on a flux of charged ions in a plasma is also being used. The development of these techniques to a useful level would not have been feasible without the reservoir of knowledge and techniques acquired through many years of research in these fields.

An electrical power source that is beginning to compete with solar cells for space applications is the radioactively heated thermionic power converter. The understanding and the techniques that make such applications possible came directly from basic studies of thermionic emission and the effect of alkali vapors on the workfunctions of metals.

The instruments used to detect charged particles, first in the Van Allen belts and now throughout space, use the techniques of scintillation, solid-state detection, and energy and mass analysis that have been developed for use with electron and ion beams in the laboratory.

Surface studies of the interaction of ion and electron beams with surfaces have led to other important industrial techniques. One of these, the process of sputtering, is widely used in the electronics industry for depositing carefully controlled films. Another is the ion implantation of doping materials into semiconductors. Indeed, it was in laboratories steeped in the techniques and traditions of such interactions of charged particles with surfaces that basic research on the behavior of electrons in the surface of semiconductors led directly to development of the transistor. Another vital space age development, the solar cell, which has supplied the power for nearly every satellite launched, arose directly from basic research on the behavior of electrons and holes at the surface of semiconductor P-N junctions.

J. HIGH-VACUUM TECHNOLOGY

Finally, few techniques have had so great an impact on industrial technology in the last two decades as the development of methods of attaining, measuring, and controlling high vacuum. The basic principles are those of atomic and molecular physics. Often the techniques are the direct products of innovative basic research in this field.

The use of directed streams of heavy atoms and molecules to carry away residual vapor in a vacuum chamber has been supplemented by ingenious applications of electrical acceleration and surface chemistry. The ultimate vacuums attainable have steadily improved with the application of research in surface physics to vacuum engineering and the use of ion pumping. Such problems as surface creep of contaminants, evolution of gases at surfaces, and reactions at ion-gauge filaments have been studied extensively in the laboratory.

Without the application of atomic and molecular physics, and surface physics in particular, to technology, the development of many commercial products and companies would not have been possible. Vacuum processing is common in the food and pharmaceutical industries. It is essential in the preparation of very pure materials, particularly those needed in the manufacture of modern electronic components. It also is used in the preparation of high-quality mirrors and other optical components.

If recent history is a valid indicator, the continued development of high-vacuum technology, made possible by research in the interactions of electrons, atoms, and molecules with each other and with surfaces, will lead to many additional new products and will constitute a major technological revolution.

12 Technological Workhorses: Metrology's Atomic Standards

It is essential to our technology-oriented nation to have a complete and consistent system of physical measurements. The system is important not only for domestic industry and commerce but also for our trade with other nations. Our system is based on atomic standards of length, time, and temperature and on the prototype standard (nonatomic) of mass.

The standards that are the basis of our measurements have been improved enormously in recent years. Today, our practical measurements involving distance, frequency, and time are linked to standards that have been created by atomic physics. These atomic standards allow far greater precision and accuracy than the prototype standards of the past. They also allow this precision and accuracy to be attained with far less effort and cost than were characteristic of the earlier, far less precise standardization methods. The commercial production of these atomic standards has made them widely available to the technological community, thus greatly enhancing their usefulness. This chapter describes the resulting economic impact of atomic standards.

Of the four basic standards, only the standard of mass is still a prototype standard–a particular "hunk" of bulk metal. Hopefully, an atomic standard of mass eventually will be used. The advent of an accurate atomic standard for mass would eliminate the fear that the size of the standard might be altered by an accident. Concerning the U.S. prototype kilogram, Allen V. Astin[10] of the National Bureau of Standards (NBS) wrote:

> The U.S. copy of the international standard of mass, known as Prototype Kilogram No. 20, is housed in a vault at the Gaithersburg laboratory of the National Bureau of Standards. It is removed no oftener than once a year for checking the values of lesser standards. Since 1889, Prototype No. 20 has been taken to France twice for comparison with the master kilogram. The national standard is never touched by hands. When it is removed from the vault, two people are always present, one to carry the kilogram in a pair of forceps, the second to catch the first if he should fall.

Although we face problems in the use of atomic standards, we do not have to worry about damaging or losing atoms. As far as we know, they do not rust, bend, shrink, or age.

The Metric Convention in Paris in 1875 was the beginning of our international agreements for standardized units of measurement. In 1960, the meter bar that was used for length standardization was finally replaced by the krypton-86 atomic wavelength standard (the meter, if defined as 1650 763.73 wavelengths of its orange-red line). Atomic wavelength measurements of fairly high precision were made before 1900, but the widespread use of atomic length standards has just begun. The methane-saturated, absorption-frequency standard of R. L. Barger and J. L. Hall (1968)[11] is a further step in the direction of highly accurate length standards. We can expect that, in turn, it will be surpassed by other even more accurate atomic standards. Probably by 1975, 100 years after the Treaty of the Meter, our best standards of length will routinely and reliably afford an accuracy one million times greater than that of the famous prototype meter bar (on those rare occasions when lesser standards are checked against it).

Hardly 15 years ago, accurate timekeeping required a combination of careful astronomical observations and fastidious maintenance of a bank of oscillators (typically quartz) over long periods of time. Today, with the hundreds of commercially produced cesium atomic beam frequency standards as well as the more elaborate but one-of-a-kind

[10] A. V. Astin, "Standards of Measurement," *Scientific American*, *218* (6), 50–62 (1968).

[11] R. L. Barger and J. L. Hall, "Pressure Shift and Broadening of Methane Line at 3.39 μ Studied by Laser-Saturated Molecular Absorption," *Phys. Rev. Lett.* *22* (1), 4–8 (1969).

laboratory cesium standards, timekeeping is easier, more accurate, and more precise by at least a factor of 10,000. The economy and performance of these atomic resonance devices continue to improve, and, as this occurs, the number of customers for these devices increases.

A scale for assigning dates is called a time scale. A scale of International Atomic Time probably will be implemented in January 1972, less than five years after the international definition of the unit of time, the second, was changed from astronomical to atomic (the second is defined as the duration of 9192 631 770 cycles of the radiation resonant, with a specified hyperfine transition of the cesium-133 atom). Most of the world's time and frequency broadcast stations will then either broadcast a stepped version (with occasional leap-seconds) of the International Atomic Time or will indicate the relevant corrections to their time as broadcast. The commonly used time scale (UTC, Universal Time Coordinated) that it will replace has been tightly tied to the errant earth. The new internationally accepted scale, with its uniform atomic rate, will allow a greater simplicity of routine operation in thousands of factories, ships, and laboratories that must do timekeeping.

Looking ahead, we see that one primary atomic standard (the choice of atom or molecule is not yet clear) soon may serve simultaneously as the most accurate standard for frequency, time, and length. And eventually the development of atomic physics may enable us to refer mass measurements also to the single, unified atomic standard–The Standard, as it would be called.

Better standards allow better measurements. But, are better measurements worthwhile? Let us consider some generalities about poor measurements.

All systems must be overdesigned (often at appreciable expense) to compensate for inadequate accuracy and precision of measurements. Not only is the expense greater, but the design time will tend to be longer, since one has to estimate the degree of overdesigning that will be adequate. W. A. Wildhack[12] of the NBS has said:

> . . . [I] naccurate measurement . . . can negate the values of extensive research . . . can spell failure for costly missiles or satellites . . . can lead to over-design, overweight, and over-cost. Always, the result is waste. . . .

The use of atomic frequency standards, though already large, is accelerating. Due in part to the unrivaled accuracy and precision of

[12] W. A. Wildhack, "Averting the Measurement Pinch," *Instrum. Soc. Am. J.*, *9* (5), 31 (1962).

frequency–time measurements that atomic frequency standards allow, there is currently a tendency to use frequency–time techniques to solve complex measurement problems. In many cases the complex problem does not involve frequency–time, but some of the proposed solutions rely on frequency–time metrology for their success.

Already more than 500 cesium-beam frequency standards have been produced in the United States, where currently three companies are marketing them. Their price is about $15,000 each. In addition, more than 1000 rubidium gas cell frequency standards have been sold; these currently are made by three companies in the United States. A complete rubidium standard sells for about $7500.

A. BROADCASTING

More than 50 atomic clocks are in use by television stations in the United States. Among other benefits, this precise control of the carrier frequencies and video timing reduces channel interference and allows accurate synchronization of frames from separated locations (no picture "roll" when switching the point of origin; also, "split-screen" operation is possible).

B. COMMUNICATIONS

Atomic standards, with their accuracy and stability, allow improvements in high-bit-rate data communications, ultranarrow-band communications, secure communications, coding techniques (for example, bipolar phase modulation), and spectrum conservation. The coherent optical radiation created by lasers, with its high carrier frequency and associated (potential) large modulation bandwidth, may give us a system having a capacity of greater than 10^{12} bits per second. To be economically justified, such a system would have to have a capacity of greater than 6×10^{10} bits per second (estimated).

A classified military system, ICNI (Integrated Communications, Navigation, Identification), with an estimated cost of greater than $1 billion, is planned. It would rely on state-of-the-art time and frequency metrology.

C. POWER DISTRIBUTION

Large parts of the U.S. electrical power system rely on standard frequency and time signals broadcast from the 60-kHz NBS station,

WWVB (controlled by atomic standards), for assistance in controlling power flow among various sections of the country.

D. LASERS

Reliable and inexpensive laser systems are produced and sold for routine length metrology (for example, one commercial unit has a usable resolution of 1 μin. together with an accuracy of better than one part in a million, for distances up to 200 ft).

At least one company is producing He–Ne lasers to sell for $48 each in lots of one thousand. Part of the mass market for this He–Ne laser is expected to be home video cartridge players.

E. NAVIGATION

The United States spends hundreds of millions of dollars on navigation each year, and the rate of expenditure is increasing. Navigation satellites probably will be an important navigation aid within the next ten years. Atomic standards will assist in the dating and synchronization of present and future navigation systems.

The two large-scale radio navigation systems, Omega and Loran C, probably will serve also as worldwide atomic time dissemination systems for a large number of users. The cost of these navigation systems will exceed $1 billion. The Omega system has only recently been approved for full worldwide implementation (eight stations) by 1972 at an estimated cost of $100 million. Equipment manufacturers believe sales of Omega user equipment will reach $4 billion once the complete system is fully operational. In addition to being a potential supplier of precise time, the entire Omega system depends heavily on the availability of state-of-the-art timing at each transmitter. All stations will be controlled by a group of atomic standards.

Similar comments apply to the Loran C navigation system. For Loran C, many more stations are involved at a cost estimated to total $250 million (ten chains). User costs will range from $100,000 downward to a few thousand dollars per receiver, depending on whether navigation or timing information is desired from the system and the degree of automation needed.

A classified military system, NAVSAT (Navigation Satellite), with an estimated cost of $2 billion, is planned. It would use state-of-the-art time-and-frequency technology.

The Air Transport Association has proposed and tested (but not yet adopted) an Aircraft Collision Avoidance System (ACAS), which also is based on advanced timing equipment and techniques. This

ACAS represents a very large potential economic impact. There would be perhaps 50 precisely controlled master ground stations, and each aircraft—possibly eventually on a worldwide basis—would have on it a precise atomic clock as part of its complex timing equipment.

Aircraft congestion is creating an urgent need for improved methods of collision avoidance. The commercially available atomic frequency standards barely achieved sufficient accuracy and stability to allow the Air Transport Association to consider seriously a time-frequency type of ACAS. If the atomic standard state of the art had not reached that critical level, a time-frequency ACAS would not have been regarded as feasible. The unit cost of the atomic standard still presents a problem in this ACAS application, but future increased mass production may solve it.

F. SPACE

The National Aeronautics and Space Administration's (NASA) immense satellite tracking networks depend heavily on precise timekeeping (atomic) at all ground stations for a great variety of space missions, including the Apollo landings on the moon. About 1200 WWVB receivers are installed at NASA tracking stations, but timing at these types of sites is so critical for both NASA and the Department of Defense that they also must use a combination of millions of dollars worth of satellites for this timing information plus expensive periodic portable clock trips to achieve and maintain adequate timing at more than 100 sites around the world. The Air Force alone spends more than $1 million each year on portable clock operations. The clocks themselves, portable and otherwise, represent a capital investment of about $1 million.

G. ASTRONOMY

Atomic standards allow highly precise and revealing astronomical measurements. As the late French astronomer, A. Danjon, suggested, atomic clocks create a vicious circle: Heretofore the motions of the stars could not be studied except by time that was itself defined by motions of the stars.[13]

Atomic devices are used in several methods of measuring the fluc-

[13] This statement is credited to Danjon by John M. Richardson. See J. M. Richardson and J. F. Brockman, "Atomic Standards of Frequency and Time," *The Phys. Teacher*, *4* (6), 247–256 (1966).

tuations in the rate of rotation of the earth. These fluctuations are still a mystery, but they might have a close connection to environmental factors such as earthquakes, shape of the earth, tides, sea level, slow movement of land masses, air circulation, heat balance of the earth and its atmosphere, and magnetic fields. One of the methods, the on-going Lunar Ranging Experiment, uses a burst of intense laser light as the measurement probe. The absolute accuracy obtained with this ranging system depends on the clocks used in the timing system. Four clocks are continuously compared against each other. These are the quartz crystal oscillator and an atomic clock on site in Texas, a tie-in to the low-frequency WWVB signal (controlled by atomic clocks), and the Loran C navigational timing signal (also controlled by atomic clocks).

4035